FUNCTIONAL SERIES

THE POCKET MATHEMATICAL LIBRARY

JACOB T. SCHWARTZ and
RICHARD A. SILVERMAN, *Editors*

PRIMERS:

1. THE COORDINATE METHOD
by I. M. Gelfand et al.

2. FUNCTIONS AND GRAPHS
by I. M. Gelfand et al.

WORKBOOKS:

1. SEQUENCES AND COMBINATORIAL PROBLEMS:
by S. I. Gelfand et al.

2. LEARN LIMITS THROUGH PROBLEMS:
by S. I. Gelfand et al.

3. MATHEMATICAL PROBLEMS: AN ANTHOLOGY
by E. B. Dynkin et al.

COURSES:

1. LIMITS AND CONTINUITY
by P. P. Korovkin

2. DIFFERENTIATION
by P. P. Korovkin

3. INFINITE SERIES: RUDIMENTS
by G. M. Fichtenholz

4. INFINITE SERIES: RAMIFICATIONS
by G. M. Fichtenholz

5. FUNCTIONAL SERIES
by G. M. Fichtenholz

FUNCTIONAL SERIES

BY

G. M. FICHTENHOLZ

English Edition
Translated and Freely Adapted by
RICHARD A. SILVERMAN

GORDON AND BREACH
SCIENCE PUBLISHERS
NEW YORK · LONDON · PARIS

Preface

This volume of The Pocket Mathematical Library concludes the study of infinite series begun in its two companion volumes *Infinite Series: Rudiments* and *Infinite Series: Ramifications*, by the same author. The latter volumes deal in detail with numerical series, i.e., infinite series whose terms are *numbers*. The present volume, on the other hand, as its name implies, is devoted to the study of infinite series whose terms are *functions*. Together all three volumes make up a comprehensive treatise on all aspects of a topic of fundamental importance in both pure and applied mathematics.

As in the companion volumes, the problems appearing at the end of each section are a vital feature of the course and cannot be neglected by the serious student, since they both illustrate and amplify the material presented in the text.

Contents

vii

CHAPTER 1

Uniform Convergence

1. Introductory Remarks

So far in the preceding two volumes of this course (see the Preface), we have considered only numerical series, i.e., series whose terms are numbers. It is true that these series sometimes involved variables as parameters, as in the case of power series (*Ruds.*, Sec. 11)[1], but in all such cases the variables were assigned constant values in the course of the discussion. For example, in proving that $\ln(1 + x)$ is the sum of the series

$$x - \frac{x^2}{2} + \frac{x^3}{3} - \cdots + (-1)^n \frac{x^n}{n} + \cdots \quad (-1 < x \leqslant 1) \qquad (1)$$

(*Ruds.*, Example 5, p. 115), we regarded x as a parameter. Thus we never took full account of the *functional* nature of the terms of a series like (1) or of the sum of such a series. It is just this aspect of the subject which will now concern us.

We begin by considering a *sequence*

$$f_1(x), f_2(x), \ldots, f_n(x), \ldots, \qquad (2)$$

whose terms are all functions of the same variable x, defined in some domain X (here X is any infinite set of numbers, which we will most often take to be an interval). Suppose this sequence has a finite limit for every x in X. Being completely determined

---

1. The abbreviation *Ruds.* refers to the volume *Infinite Series: Rudiments* by G. M. Fichtenholz, (The Pocket Mathematical Library).

by the value of x, this limit is itself a function

$$f(x) = \lim_{n \to \infty} f_n(x) \qquad (3)$$

of x (in X), which we call the *limit function* of the sequence (2), or of the (variable) function $f_n(x)$.

We are now interested not merely in the existence of the limit (3) for each separate value of x, but also in the properties of this limit *regarded as a function of x*. To illustrate the new kind of problems to which this point of view gives rise, suppose the terms of the sequence (2) are all continuous functions of x in some interval X. Then does this guarantee the continuity of the limit function? Not at all! In fact, the limit function may or may not be a continuous function itself, as shown by the following examples:

Example 1. If $f_n(x) = x^n$, $X = [0, 1]$, then

$$f(x) = \begin{cases} 0 \text{ if } x < 1, \\ 1 \text{ if } x = 1, \end{cases}$$

so that $f(x)$ is discontinuous at $x = 1$.

Example 2. If

$$f_n(x) = \frac{1}{1 + nx},$$

$X = [0, \infty)$, then

$$f(x) = \begin{cases} 0 \text{ if } x > 0, \\ 1 \text{ if } x = 0, \end{cases}$$

so that $f(x)$ it discontinuous at $x = 0$.

Example 3. If

$$f_n(x) = \frac{nx}{1 + n^2 x^2},$$

$X = (-\infty, \infty)$, then $f(x) = 0$ for all x, so that $f(x)$ is continuous everywhere.

Thus the problem naturally arises of finding conditions under which the limit function will be continuous, and this will be done in Sec. 4.

As already noted (*Ruds.*, Sec. 1), the study of a numerical series and its sum is just another way of studying a numerical sequence and its limit. The same applies to the "functional series"

$$\sum_{n=1}^{\infty} u_n(x) = u_1(x) + u_2(x) + \cdots + u_n(x) + \cdots, \qquad (4)$$

whose terms are all functions of the same variable x, defined in some domain X. Suppose the series (4) converges for every x in X. Then its sum is itself a function $f(x)$ of the variable x, in fact the function defined by (3) if we take $f_n(x)$ to be the *partial sum*

$$f_n(x) = u_1(x) + u_2(x) + \cdots + u_n(x).$$

Conversely, the problem of determining the limit function of any given sequence (2) can be regarded as the problem of summing a series of the form (4), if we set

$$u_1(x) = f_1(x), \quad u_2(x) = f_2(x) - f_1(x), \ldots,$$

$$u_n(x) = f_n(x) - f_{n-1}(x), \ldots$$

Most of the time, we will deal with functional series of the form (4) rather sequences of functions, since this version of the problem of studying the limit function is usually more convenient.

We emphasize once again that this volume will be concerned not only with questions of convergence for a series of the form (4), but also with the study of "functional properties" of the sum of such a series. In this regard, we again cite the problem of whether or not the sum of a given functional series with continuous terms is itself a continuous function.

It turns out that the functional properties of the limit function of a sequence (or, equivalently, of the sum function of a series)

depends in an essential way on the character of the approximation of $f_n(x)$ to $f(x)$ for various values of x. Here various pos-
sibilities can occur, as discussed in the next section.

PROBLEM

Is the limit of the sequence

$$f_n(x) = 2n^2 x \, e^{-n^2 x^2} \quad (-\infty < x < \infty)$$

a continuous function?

2. Uniform and Nonuniform Convergence

Given a sequence of functions

$$f_1(x), f_2(x), \ldots, f_n(x). \tag{1}$$

defined in some set X, suppose

$$f(x) = \lim_{n \to \infty} f_n(x) \tag{2}$$

for all x in X. Then, by the very definition of a limit, this means
that *given any fixed x in X* (so that (1) becomes a numerical
sequence) and any positive number ε, there is an integer $N > 0$
such that

$$|f_n(x) - f(x)| < \varepsilon \tag{3}$$

for all $n > N$, *where the value of x figuring in (3) is again the
given fixed value.* For *another* fixed value of x, (2) gives another
numerical sequence. Then for the same $\varepsilon > 0$, the number N
may no longer be suitable and may have to be replaced by a
larger number. But x takes *infinitely many* values, so that we
are in effect dealing with *infinitely many* different convergent
numerical sequences, each with its own value of N. The question
now arises of whether or not there exists a number N (for a
given ε) which is suitable for all these sequences *simultaneously.*

Such a number N may or may not exist, as shown by the following examples:

Example 1. Let

$$f_n(x) = \frac{x}{1 + n^2x^2} \quad (0 \leqslant x \leqslant 1), \tag{4}$$

so that

$$f(x) = \lim_{n \to \infty} f_n(x) = 0 \quad (0 \leqslant x \leqslant 1).$$

Since

$$0 \leqslant f_n(x) = \frac{1}{2n} \frac{2nx}{1 + n^2x^2} \leqslant \frac{1}{2n},$$

it is immediately clear that the inequality $f_n(x) < \varepsilon$ holds for all $n > 1/2\varepsilon$ *regardless of the value of x in the interval* [0, 1]. Thus, for example, the number[2]

$$N = \left[\frac{1}{2\varepsilon}\right]$$

is suitable *for all x simultaneously.*

Example 2. Next let

$$f_n(x) = \frac{nx}{1 + n^2x^2} \quad (0 \leqslant x \leqslant 1) \tag{5}$$

(cf. Example 3, p. 2). In this case,

$$f(x) = \lim_{n \to \infty} f_n(x) = 0 \quad (0 \leqslant x \leqslant 1).$$

For any fixed $x > 0$, we need only choose

$$n > \left[\frac{1}{x\varepsilon}\right]$$

to make

$$f_n(x) < \frac{1}{nx} < \varepsilon.$$

---

2. By [a] is meant the *integral part* of a, i.e., the largest integer $\leqslant a$.

But, on the other hand, no matter how large n is chosen, we can always find a point in the interval $[0, 1]$, namely the point $x = 1/n$, at which

$$f_n(x) = f_n\left(\frac{1}{n}\right) = \frac{1}{2}.$$

Thus there is no way of choosing n large enough to make $f_n(x) < \frac{1}{2}$ *for all values of x in the interval* $[0, 1]$. In other words, even for ε as large as $\frac{1}{2}$, there is no number N suitable for all x simultaneously.

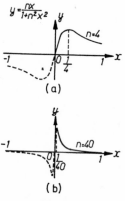

Figure 1

We now interpret these examples graphically. Figure 1 shows the graphs of the function (5) for $n = 4$ and $n = 40$. Note the characteristic peak of height $\frac{1}{2}$ which moves to the left as n increases. Although as n increases, the points of successive curves approach the x-axis along every vertical line, taken separately, there is no curve which *as a whole* "hugs" the x-axis along the entire interval $0 \leqslant x \leqslant 1$.

The situation is entirely different for the function (4) considered in Example 1. There is no need to draw the corresponding graphs, since they can easily be deduced from those of the function (5). For example, the graph of (4) for $n = 4$ is obtained from the curve in Figure 1a by reducing all ordinates 4 times, while the graph of (4) for $n = 40$ is obtained from the curve in Figure 1b by reducing all ordinates 40 times. Note that in this case, the curves eventually "hug" the x-axis along the entire interval $0 \leqslant x \leqslant 1$.

The above considerations suggest

DEFINITION 1. *Let*

$$f_1(x), f_2(x), \ldots, f_n(x), \ldots \quad (x \ in \ X)$$

be a sequence of functions converging for all x in X to a limit function $f(x)$. Suppose that given any $\varepsilon > 0$, there is an integer $N > 0$ (**independent of x**) *such that the inequality*

$$|f_n(x) - f(x)| < \varepsilon$$

holds **for all x in X** *whenever $n > N$. Then the sequence is said to* **converge uniformly** *(in X) to the function $f(x)$.*[3]

Thus in Example 1 the function $f_n(x)$ approaches zero uniformly for all x in [0, 1]. However, in Example 2 the convergence of $f_n(x)$ to zero is *nonuniform.*

Example 3. Let

$$f_n(x) = x^n \quad (0 \leqslant x \leqslant 1),$$

as in Example 1, p. 2. Then the inequality $x^n < \varepsilon$ $(\varepsilon < 1)$ cannot be satisfied *for all $x < 1$ simultaneously*, since $x^n \to 1$ (for fixed n) as $x \to 1$. Hence $f_n(x)$ does not approach its limit function

$$f(x) = \lim_{n \to \infty} f_n(x) = \begin{cases} 0 & \text{if } x < 1, \\ 1 & \text{if } x = 1 \end{cases}$$

uniformly. This failure of "uniformity" is apparent from Figure 2, where the graph of $f_n(x)$ rises to a peak at the same point $x = 1$ for every n.

---

3. In this case, we also say that the (variable) function $f_n(x)$ *approaches* $f(x)$ *uniformly.*

The failure of uniformity can also be seen by noting that the inequality $x^n < \varepsilon$ is equivalent to

$$n > \frac{\ln \varepsilon}{\ln x} \quad (0 < x < 1, 0 < \varepsilon < 1). \tag{6}$$

But the expression on the right approaches infinity as $x \to 1$, and hence no number n can satisfy the inequality (6) for all values of x.

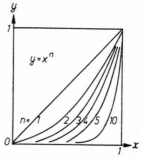

Figure 2

We now paraphrase everything just said about uniform convergence of a sequence of functions for the case of a functional series

$$\sum_{n=1}^{\infty} u_n(x) = u_1(x) + u_2(x) + \cdots + u_n(x) + \cdots, \tag{7}$$

whose terms are defined for all x in some set X. Assuming that the series (7) is convergent for all x in X, let $f(x)$ be the sum of (7), let

$$f_n(x) = u_1(x) + u_2(x) + \cdots + u_n(x)$$

be its *nth partial sum,* and let

$$\varphi_n(x) = \sum_{k=n+1}^{\infty} u_k(x) = f(x) - f_n(x)$$

be its *remainder* after n terms. Then

$$\lim_{n \to \infty} f_n(x) = f(x), \quad \lim_{n \to \infty} \varphi_n(x) = 0$$

for any fixed x. *If the partial sum $f_n(x)$ approaches the sum $f(x)$* **uniformly** *(or equivalently, if the remainder $\varphi_n(x)$ approaches zero uniformly), for all x in X, then the series* (7) *is said to converge to $f(x)$* **uniformly** *(in X).* This definition is clearly equivalent to

DEFINITION 2. *Let*

$$\sum_{n=1}^{\infty} u_n(x) = u_1(x) + u_2(x) + \cdots + u_n(x) + \cdots \quad (x \text{ in } X)$$

be a functional series converging for all x in X to a sum function $f(x)$. Suppose that given any $\varepsilon > 0$, there is an integer $N > 0$ **(independent of x)** *such that the inequality*

$$|f_n(x) - f(x)| = \left| \sum_{k=1}^{n} u_k(x) - f(x) \right| < \varepsilon$$

holds **for all x in X** *whenever $n > N$. Then the series is* **said to converge uniformly** *(in X) to the function $f(x)$.*

Naturally, we can construct examples of uniformly and non-uniformly convergent series by transcribing our previous examples (involving sequences) in the language of series. Besides these examples, we now give two others:

Example 4. Consider the series

$$\sum_{n=1}^{\infty} x^{n-1}, \tag{8}$$

which converges in the interval $X = (-1, 1)$. For any x in X the remainder after n terms is

$$\varphi_n(x) = \frac{x^n}{1 - x}.$$

For every *fixed n*, we have

$$\lim_{x \to -1+} |\varphi_n(x)| = \frac{1}{2}, \qquad \lim_{x \to 1-} \varphi_n(x) = \infty.$$

Either of these formulas shows that it is impossible to satisfy the inequality

$$|\varphi_n(x)| < \varepsilon \qquad \left(\varepsilon < \frac{1}{2} \right)$$

for all x simultaneously for the same integer n. Thus the convergence of the series (8) in the interval $(-1, 1)$ is *nonuniform*, and the same is true separately of the intervals $(-1, 0]$ and $[0, 1)$.

Example 5. The series

$$\sum_{n=1}^{\infty} \frac{(-1)^{n-1}}{x^2 + n} \tag{9}$$

converges for every x in the infinite interval $X = (-\infty, \infty)$, since it satisfies the conditions of Leibniz's test (*Ruds.*, p. 73). Moreover, the absolute value of the remainder $\varphi_n(x)$ of the series can be estimated by its first term, i.e.,

$$|\varphi_n(x)| < \frac{1}{x^2 + n + 1} \leqslant \frac{1}{n + 1}$$

(*Ruds.*, p. 74). If follows that the series (9) converges *uniformly* in the whole interval $(-\infty, \infty)$.

PROBLEMS

1. Show that neither of the functions

$$f_n(x) = \frac{1}{1 + nx} \quad (0 \leqslant x \leqslant 1)$$

or

$$g_n(x) = 2n^2 x\, e^{-n^2 x^2} \quad (0 \leqslant x \leqslant 1)$$

approaches its limit uniformly.

Hint. Note that

$$f_n\left(\frac{1}{n}\right) = \frac{1}{2}, \qquad g_n\left(\frac{1}{n}\right) = \frac{2n}{e}.$$

Also note that

$$\frac{1}{1 + nx} < \varepsilon$$

is equivalent to

$$n > \frac{1}{x}\left(\frac{1}{\varepsilon} - 1\right) \quad (0 < x < 1, 0 < \varepsilon < 1),$$

where the expression on the right approaches infinity as $x \to 0$.

2. Prove that the series

$$\sum_{n=1}^{\infty} \frac{(-1)^{n-1}x^2}{(1 + x^2)^n} \tag{10}$$

converges uniformly in the interval $(-\infty, \infty)$.

Hint. Note that

$$|\varphi_n(x)| < \frac{x^2}{(1 + x^2)^n} = \frac{x^2}{1 + nx^2 + \cdots} < \frac{1}{n} \quad (x \neq 0).$$

3. Prove that the series

$$\sum_{n=1}^{\infty} \frac{x^2}{(1 + x^2)^n},$$

obtained by taking absolute values of the terms of (10), converges in the interval $(-\infty, \infty)$, but not uniformly.

Hint. For any fixed n,

$$\varphi_n(x) = \frac{\dfrac{x^2}{(1 + x^2)^{n+1}}}{1 - \dfrac{1}{1 + x^2}} = \frac{1}{(1 + x^2)^n} \quad (x \neq 0)$$

approaches 1 as $x \to 0$.

4. Prove that the convergence of the sequence of functions (5) is uniform in every interval $[a, 1]$, where $0 < a < 1$, but non-uniform in every interval $[0, a]$.

Hint. For all $x \geqslant a$,

$$f_n(x) = \frac{nx}{1 + n^2x^2} \leqslant \frac{n}{1 + n^2a^2} < \frac{1}{na^2}.$$

Comment. Thus the property of nonuniformity "condenses," as it were, about the point $x = 0$, which we call a *point of non-uniformity*. The same role is played by the point $x = 1$ in Example 3 and by both points $x = \pm 1$ in Example 4. In more complicated cases, there may be *infinitely many* points of non-uniformity.

3. Tests for Uniform Convergence

The Cauchy convergence criterion for numerical series (*Ruds.*, p. 61) leads to the following criterion for *uniform* convergence of functional series:

THEOREM 1. *The series*

$$\sum_{n=1}^{\infty} u_n(x) = u_1(x) + u_2(x) + \cdots + u_n(x) + \cdots \quad (x \text{ in } X) \quad (1)$$

is uniformly convergent in the set X if and only if, given any $\varepsilon > 0$, there is a positive integer N such that, for every x in X, the inequality

$$\left| \sum_{k=n+1}^{n+m} u_k(x) \right| = |u_{n+1}(x) + u_{n+2}(x) + \cdots + u_{n+m}(x)| < \varepsilon \quad (2)$$

holds for all $n > N$ and $m = 1, 2, \ldots$

Proof. Suppose the series (1) is uniformly convergent, with sum $f(x)$. Then the sequence of partial sums

$$f_n(x) = \sum_{k=1}^{n} u_k(x) = u_1(x) + u_2(x) + \cdots + u_n(x)$$

converges uniformly to $f(x)$ in X. Therefore, by Definition 1, p. 7, given any $\varepsilon > 0$, there is an integer $N > 0$ (independent of x) such that

$$|f_n(x) - f(x)| < \frac{\varepsilon}{2} \quad (3)$$

for all x in X whenever $n > N$, and hence, in particular,

$$|f_{n+m}(x) - f(x)| < \frac{\varepsilon}{2} \quad (m = 1, 2, \ldots). \tag{4}$$

It follows from (3) and (4) that, for every x in X,

$$|f_{n+m}(x) - f_n(x)| = \left| \sum_{k=1}^{n+m} u_k(x) - \sum_{k=1}^{n} u_k(x) \right| = \left| \sum_{k=n+1}^{n+m} u_k(x) \right| < \varepsilon$$

for all $n > N$ and $m = 1, 2, \ldots$

Conversely, suppose that given any $\varepsilon > 0$, there is an integer $N > 0$ such that (2) holds for every x in X and all $n > N$, $m = 1, 2, \ldots$ Then (1) converges for every x in X, by the Cauchy convergence criterion for numerical series. Let $f(x)$ denote the sum of the series (1). Given any $\varepsilon > 0$, let $0 < \varepsilon' < \varepsilon$ and choose N such that

$$\left| \sum_{k=1}^{n} u_k(x) - \sum_{k=1}^{n+m} u_k(x) \right| = \left| \sum_{k=1}^{n+m} u_k(x) - \sum_{k=1}^{n} u_k(x) \right|$$

$$= \left| \sum_{k=n+1}^{n+m} u_k(x) \right| < \varepsilon'$$

for all x in X and all $n > N$, $m = 1, 2, \ldots$ Taking the limit of the left-hand side as $m \to \infty$, we get

$$\left| \sum_{k=1}^{n} u_k(x) - f(x) \right| \leqslant \varepsilon' < \varepsilon$$

for all x in X whenever $n > N$. Therefore the series (1) is uniformly convergent, by Definition 2, p. 9. ∎[4]

COROLLARY 1. *The sequence*

$$f_1(x), f_2(x), \ldots, f_n(x), \ldots \quad (x \text{ in } X)$$

is uniformly convergent in the set X if and only if, given any $\varepsilon > 0$, there is a positive integer N such that, for every x in X, the inequality $\quad |f_{n+m}(x) - f_n(x)| < \varepsilon$

holds for all $n > N$ and $m = 1, 2, \ldots$

---

4. The symbol ∎ stands for Q.E.D. and indicates the end of a proof.

Proof. Apply Theorem 1 to the series

$$f_1(x) + [f_2(x) - f_1(x)] + \cdots + [f_n(x) - f_{n-1}(x)] + \cdots. \quad \blacksquare$$

COROLLARY 2. *If every term of a series* (1) *uniformly convergent in a set* X *is multiplied by a function bounded in* X, *i.e., a function* $v(x)$ *such that*

$$|v(x)| \leqslant M$$

for all x *in* X, *then the resulting series is still uniformly convergent in* X.

Proof. Given any $\varepsilon > 0$, choose N such that

$$\left| \sum_{k=n+1}^{n+m} u_k(x) \right| < \frac{\varepsilon}{M} \quad (x \text{ in } X)$$

for all $n > N$ and $m = 1, 2, \ldots$ This is possible by Theorem 1. Then

$$\left| \sum_{k=n+1}^{n+m} u_k(x)\, v(x) \right| \leqslant |v(x)| \sum_{k=1}^{n+m} u_k(x) < M \frac{\varepsilon}{M} = \varepsilon \quad (x \text{ in } X)$$

for all $n > N$ and $m = 1, 2, \ldots$ Now apply Theorem 1 again. $\quad \blacksquare$

Theorem 1 is not a very practical tool for establishing the uniform convergence of a given functional series. However, starting from Theorem 1, we can easily establish a number of convenient tests for uniform convergence:

THEOREM 2 **(Weierstrass' test).** *Suppose the terms of a functional series*

$$\sum_{n=1}^{\infty} u_n(x) \quad (x \text{ in } X) \tag{5}$$

satisfy the inequalities[5]

$$|u_n(x)| \leqslant c_n \quad (n = 1, 2, \ldots) \tag{6}$$

---

5. If (6) holds, we say that the series (5) is *majorized* by the series (7), or that (7) is a *majorant* series of (5).

for all x in X, where the numbers c_n are the terms of a convergent numerical series

$$\sum_{n=1}^{\infty} c_n. \tag{7}$$

Then the series (5) is uniformly convergent in X.

Proof. It follows from (6) that

$$|u_{n+1}(x) + u_{n+2}(x) + \cdots + u_{n+m}(x)|$$

$$\leqslant c_{n+1} + c_{n+2} + \cdots + c_{n+m} \tag{8}$$

for all x in X. By the Cauchy convergence criterion for numerical series, given any $\varepsilon > 0$, there is an integer $N > 0$ such that the right-hand side of (8) is less than ε for all $n > N$ and $m = 1, 2, \ldots$ The uniform convergence of (5) is now an immediate consequence of Theorem 1. ∎

Example 1. If the series

$$\sum_{n=1}^{\infty} a_n$$

is *absolutely* convergent, then the series

$$\sum_{n=1}^{\infty} a_n \cos nx, \quad \sum_{n=1}^{\infty} a_n \sin nx \tag{9}$$

converge uniformly in any interval. In fact,

$$|a_n \sin nx| \leqslant |a_n|, \quad |a_n \cos nx| \leqslant |a_n|,$$

so that the series (9) are majorized by the series

$$\sum_{n=1}^{\infty} |a_n|.$$

Next we prove two tests involving functional series of the form

$$\sum_{n=1}^{\infty} a_n(x) \, b_n(x) = a_1(x) \, b_1(x) + a_2(x) \, b_2(x) + \cdots$$

$$+ a_n(x) \, b_n(x) + \cdots, \tag{10}$$

where $a_n(x)$ and $b_n(x)$ $(n = 1, 2, ...)$ are functions of x in some set X. These tests are the exact analogues of the tests of Abel and Dirichlet (*Ruds.*, Sec. 13), and hence will be designated by the same names.

THEOREM 3 **(Abel's test).** *Suppose that*
a) *The sequence* $\{a_n(x)\}$ *is nonincreasing (or nondecreasing) for every fixed x and is uniformly bounded in X,[6] i.e.,*

$$|a_n(x)| \leqslant K \quad (x \text{ in } X, n = 1, 2, ...)$$

for some constant K;
b) *The series*

$$\sum_{n=1}^{\infty} b_n(x) = b_1(x) + b_2(x) + \cdots + b_n(x) + \cdots \quad (11)$$

converges uniformly in X.
Then the series (10) *also converges uniformly in X.*

Proof. It follows from Theorem 1 that, given any $\varepsilon > 0$, there is an integer $N > 0$ such that

$$\left| \sum_{k=n+1}^{n=m} b_k(x) \right| < \varepsilon$$

for all x in X and $n > N$, $m = 1, 2, ...$ Hence, by Abel's lemma (*Ruds.*, p. 78).

$$\left| \sum_{k=n+1}^{n+m} a_k(x)\, b_k(x) \right| \leqslant \varepsilon\, (|a_{n+1}(x) + 2\, |a_{n+m}(x)||) \leqslant 3K\varepsilon$$

for all x in X and $n > N$, $m = 1, 2, ...$ The uniform convergence of (10) in X now follows by using Theorem 1 again. ■

THEOREM 4 **(Dirichlet's test).** *Suppose that*
a) *The sequence* $\{a_n(x)\}$ *is nonincreasing (or nondecreasing) for every fixed x and converges to zero uniformly in X;*
b) *The partial sums*

$$B_n(x) = b_1(x) + b_2(x) + \cdots + b_n(x) \quad (n = 1, 2, ...)$$

---

6. Here the word "uniformly" refers to the fact that the inequality $|a_n(x)| \leqslant K$ holds for all x in X as well as for all $n = 1, 2, ...$

of the series (11) *form a sequence uniformly bounded in X,*
i.e.,

$$|B_n(x)| \leqslant K \quad (x \text{ in } X, n = 1, 2, \ldots)$$

for some constant K.
Then the series (10) *converges uniformly in X.*

Proof. Given any $\varepsilon > 0$, there is an integer $N > 0$ such that $|a_n(x)| < \varepsilon$ for all x in X and $n > N$. Moreover,

$$|b_{n+1}(x) + b_{n+2}(x) + \cdots + b_{n+p}(x)| = |B_{n+p}(x) - B_n(x)| \leqslant 2K$$

for all x in X and $p = 1, 2, \ldots$ It follows from Abel's lemma that

$$\left| \sum_{k=n+1}^{n+m} a_k(x) \, b_k(x) \right| \leqslant 2L \left(|a_{n+1}(x)| + 2 \, |a_{n+m}(x)| \right) \leqslant 6K\varepsilon$$

for all x in X and $m = 1, 2, \ldots$ The uniform convergence of (10) in X is now an immediate consequence of Theorem 1. ∎

Remark. In practice, one often encounters an ordinary numerical sequence $\{a_n\}$ instead of the sequence of functions $\{a_n(x)\}$, or an ordinary numerical series

$$\sum_{n=1}^{\infty} b_n$$

instead of the functional series (11). These are obviously special cases of the more general case treated above, since a convergent numerical sequence or convergent numerical series can be regarded as uniformly convergent (there being no dependence on x at all).

PROBLEMS

1. Suppose the series

$$\sum_{n=1}^{\infty} u_n(x) \tag{12}$$

is uniformly convergent in X. Show that the terms of (12) can be grouped in such a way that the resulting series satisfies the conditions of Weierstrass' test.

Hint. Given any convergent positive series

$$\sum_{n=1}^{\infty} c_n,$$

find an integer m_1 such that

$$|u_{m_1+1}(x) + \cdots + u_n(x)| < c_1$$

for all x in X and $n > m_1$, then an integer m_2 such that

$$|u_{m_2+1}(x) + \cdots + u_n(x)| < c_2$$

for all x in X and $n > m_2$, and so on. Group the terms of (12) as follows:

$$[u_1(x) + \cdots + u_{m_1}(x)] + [u_{m_1+1}(x) + \cdots + u_{m_2}(x)]$$
$$+ [u_{m_2+1}(x) + \cdots + u_{m_3}(x)] + \cdots.$$

Then note that

$$|u_{m_k+1}(x) + \cdots + u_{m_{k+1}}(x)| \leqslant c_k \quad (k = 1, \ldots)$$

for all x in X.

2. Prove that if Weierstrass' test is applicable to a series (12), then the series necessarily converges absolutely and moreover the series

$$\sum_{n=1}^{\infty} |u_n(x)|, \tag{13}$$

whose terms are the absolute values of those of (12), is also uniformly convergent in X.

3. Give an example where the series (12) converges uniformly but not absolutely.

Hint. The series (9), p. 10 is not absolutely convergent, since

$$\left| \frac{(-1)^{n+1}}{x^2 + n} \right| > \frac{1}{2n}$$

for all sufficiently large n.

4. Give an example where the series (12) converges absolutely and uniformly, but the series (13) is not uniformly convergent. *Hint.* See Problems 2 and 3, p. 11.

5. Let $\{a_n\}$ be a decreasing sequence converging to zero. Prove that the series

$$\sum_{n=1}^{\infty} a_n \sin nx, \quad \sum_{n=1}^{\infty} a_n \cos nx$$

are uniformly convergent in any closed interval X which does not contain points of the form $2k\pi$ ($k = 0, \pm 1, \pm 2, \ldots$).

Hint. Use Dirichlet's test, noting for example that

$$\left| \sum_{k=1}^{n} \sin kx \right| = \left| \frac{\cos \dfrac{x}{2} - \cos\left(n + \dfrac{1}{2}\right)x}{2 \sin \dfrac{x}{2}} \right| \leqslant \frac{1}{\left| \sin \dfrac{x}{2} \right|}$$

(*Ruds.*, p. 81), where the expression on the right is bounded from above in any interval X of the indicated type.

CHAPTER 2

Functional Properties of the Sum of a Series

4. Continuity of the Sum of a Series

We now study the relation between the functional properties of the sum of a series whose terms are functions and the functional properties of the terms themselves. Functional series are encountered in practice much more often than sequences of functions, and hence our subsequent discussion will be based on a series point of view. However, as already noted, the series point of view and the sequence point of view are entirely equivalent, and it will be an easy matter to transcribe all results involving series into the language of sequences (see Problems 7–10, pp. 41–43). The concept of uniform convergence introduced in Chapter 1 will play a decisive role in all that follows, thereby revealing the full importance of the concept.

THEOREM 1. *Let*

$$\sum_{n=1}^{\infty} u_n(x) = u_1(x) + u_2(x) + \cdots + u_n(x) + \cdots \tag{1}$$

be a functional series, each of whose terms $u_n(x)$ is defined in an interval $X = [a, b]$ and continuous at a point $x = x_0$ in X. Suppose the series (1) converges uniformly in X to a function $f(x)$. Then $f(x)$ is also continuous at $x = x_0$.

Proof. Let $f_n(x)$ be the nth partial sum of (1) and $\varphi_n(x)$ its remainder after n terms, as in Sec. 1. Then

$$f(x) = f_n(x) + \varphi_n(x), \tag{2}$$

20

and in particular

$$f(x_0) = f_n(x_0) + \varphi_n(x_0),$$

so that

$$|f(x) - f(x_0)| \leqslant |f_n(x) - f_n(x_0)| + |\varphi_n(x) - \varphi_n(x_0)|$$

$$\leqslant |f_n(x) - f_n(x_0)| + |\varphi_n(x)| + |\varphi_n(x_0)|. \quad (3)$$

Given any $\varepsilon > 0$, there is an integer n such that

$$|\varphi_n(x)| < \varepsilon \quad (4)$$

for all x in X (including the point $x = x_0$), because of the uniform convergence of (1). But for every fixed n, the function $f_n(x)$ is the sum of a finite number of functions $u_k(x)$, each continuous at the point $x = x_0$, and hence $f_n(x)$ itself is continuous at the point $x = x_0$. Thus there is a number $\delta > 0$ such that

$$|f_n(x) - f_n(x_0)| < \varepsilon \quad (5)$$

whenever $|x - x_0| < \delta$. Combining (3), (4) and (5), we find that

$$|f(x) - f(x_0)| < 3\varepsilon$$

whenever $|x - x_0| < \delta$, i.e., $f(x)$ is continuous at $x = x_0$. ∎

Remark. In particular, if the functions $u_n(x)$ are continuous in the whole interval $X = [a, b]$ and if the series (1) is uniformly convergent in X, then the sum $f(x)$ is continuous in X. In fact, we need only note that being continuous *in X* means being continuous *at every point of X*.

The requirement of *uniform* convergence cannot be dropped in Theorem 1 (see Problem 1). In general, uniform convergence is only a *sufficient* condition for continuity of the sum of (1) and not a *necessary* condition (see Problem 2). On the other hand, if the terms of (1) are *nonnegative*, then uniform convergence turns out to be a necessary as well as a sufficient condition for continuity of the sum of (1):

THEOREM 2 **(Dini).** *Let* (1) *be a functional series, each of whose terms $u_n(x)$ is continuous and nonnegative in the whole interval*

$X = [a, b]$. *Suppose the series* (1) *converges in* X *to a function* $f(x)$ *which is also continuous in the whole interval* X. *Then the series is uniformly convergent in* X.

Proof. The remainder

$$\varphi_n(x) = \sum_{k=n+1}^{\infty} u_k(x) = f(x) - f_n(x)$$

of the series (1), being the difference between two continuous functions, is itself continuous. Since the terms of the series are nonnegative, the sequence $\{\varphi_n(x)\}$ is nonincreasing for every fixed x:

$$\varphi_1(x) \geqslant \varphi_2(x) \geqslant \cdots \geqslant \varphi_n(x) \geqslant \varphi_{n+1}(x) \geqslant \cdots.$$

Moreover,
$$\lim_{n \to \infty} \varphi_n(x) = 0$$

for every fixed x, since the series (1) is convergent in X. To prove the uniform convergence of the series, we need only show that given any $\varepsilon > 0$, there exists at least one value of n such that $\varphi_n(x) < \varepsilon$ for all x, since the inequality will then automatically hold for all larger values of n. We now prove this by contradiction. Suppose such a number n does not exist for some $\varepsilon > 0$. Then for every $n = 1, 2, \ldots$ there is a point $x = x_n$ in the interval X such that $\varphi_n(x_n) \geqslant \varepsilon$. The sequence $\{x_n\}$ is contained in the finite interval $X = [a, b]$, and hence is bounded. It follows from the Bolzano–Weierstrass theorem[1] that the sequence $\{x_n\}$ has a convergent subsequence $\{x_{n_k}\}$, with limit x_0, say. Since $\varphi_m(x)$ is continuous, we have

$$\lim_{k \to \infty} \varphi_m(x_{n_k}) = \varphi_m(x_0)$$

for every m. On the other hand, given any m, $n_k \geqslant m$ for sufficiently large k, and hence

$$\varphi_m(x_{n_k}) \geqslant \varphi_{n_k}(x_{n_k}) \geqslant \varepsilon$$

---

1. See e.g., P.P. Korovkin, *Limits and Continuity* (in The Pocket Mathematical Library), Corollary 1, p. 60.

for sufficiently large k. Taking the limit as $k \to \infty$, we find that

$$\lim_{k \to \infty} \varphi_m(x_{n_k}) = \varphi_m(x_0) \geqslant \varepsilon,$$

contrary to the fact that

$$\lim_{m \to \infty} \varphi_m(x_0) = 0. \quad \blacksquare$$

We now prove a generalization of Theorem 1, involving passing to the limit "term by term" in a functional series:

THEOREM 3. *Let X be any infinite set with a (finite or infinite) limit point a,[2] and let* (1) *be a functional series, each of whose terms $u_n(x)$ is defined in X and has a finite limit as x approaches a:*

$$\lim_{x \to a} u_n(x) = c_n \quad (n = 1, 2, \ldots). \tag{6}$$

Suppose the series (1) *converges uniformly in X to a function $f(x)$. Then*

1) *The numerical series*

$$\sum_{n=1}^{\infty} c_n \tag{7}$$

made up of the limits (6) *converges to a sum C;*

2) *The sum of the series* (1) *approaches C as $x \to a$:*

$$\lim_{x \to a} f(x) = C. \tag{8}$$

Proof. By Theorem 1, p. 12, given any $\varepsilon > 0$, there is an integer $N > 0$ such that

$$\left| \sum_{k=n+1}^{n+m} u_k(x) \right| = |u_{n+1}(x) + u_{n+2}(x) + \cdots + u_{n+m}(x)| < \varepsilon$$

for all $n > N$ and $m = 1, 2, \ldots$ Taking the limit as $x \to a$ and using (6), we get

$$|c_{n+1} + c_{n+2} + \cdots + c_{n+m}| \leqslant \varepsilon.$$

---

2. The point a itself need not belong to X.

Hence the series (7) converges, by the Cauchy convergence principle. Let C, C_n and γ_n denote the sum, nth partial sum and remainder after n terms of (7), so that

$$C = C_n + \gamma_n. \tag{9}$$

Subtracting (9) from (2), we find that

$$|f(x) - C| \leqslant |f_n(x) - C_n| + |\varphi_n(x)| + |\gamma_n|. \tag{10}$$

By the uniform convergence of (1) and the convergence of (7), given any $\varepsilon > 0$, we can find an n so large that

$$|\varphi_n(x)| < \varepsilon \tag{11}$$

for all x in X, and also

$$|\gamma_n| < \varepsilon. \tag{12}$$

But obviously

$$\lim_{x \to a} f_n(x) = \lim_{x \to a} \sum_{k=1}^{n} u_k(x) = \sum_{k=1}^{n} c_k = C_n.$$

Hence, for the case of finite a, there is a $\delta > 0$ such that

$$|f_n(x) - C_n| < \varepsilon \tag{13}$$

whenever $|x - a| < \delta$ (the case of infinite a is treated similarly). Combining (10)–(13), we find that

$$|f(x) - C| < 3\varepsilon$$

whenever $|x - a| < \delta$. ■

Remark 1. Note the resemblance between this proof and that of Theorem 1.

Remark 2. Writing (8) in the expanded form

$$\lim_{x \to a} \sum_{n=1}^{\infty} u_n(x) = \sum_{n=1}^{\infty} \left\{ \lim_{x \to a} u_n(x) \right\},$$

we can express Theorem 3 as follows: If a series is uniformly convergent, then the limit of the sum of the series equals the sum of the series made up of the limits of its terms, i.e., the limit of such a series can be taken "term by term."

PROBLEMS

1. Give an example of a convergent series of continuous functions with a discontinuous sum.

Ans. For example, the *nonuniformly* convergent series

$$\sum_{n=1}^{\infty} \frac{x^2}{(1 + x^2)^n},$$

considered in Problem 3, p. 11, has the discontinuous sum

$$f(x) = \begin{cases} 1 & \text{if} \quad x \neq 0, \\ 0 & \text{if} \quad x = 0. \end{cases}$$

2. Show that each of the series

$$\sum_{n=1}^{\infty} \left[\frac{nx}{1 + n^2 x^2} - \frac{(n-1)x}{1 + (n-1)^2 x^2} \right]$$

and

$$\sum_{n=1}^{\infty} 2x \left[n^2 e^{-n^2 x^2} - (n-1)^2 e^{-(n-1)^2 x^2} \right]$$

has a continuous sum (identically equal to zero) in the interval [0, 1], although the convergence is nonuniform in both cases.

Hint. Cf. Example 2, p. 5 and Problem 1, p. 10.

3. Show that the condition of uniform convergence in Theorem 1 can be replaced by the following weaker condition: Given any $\varepsilon > 0$ and any integer $N > 0$, there is *at least one* integer $n > N$ such that

$$\left| \sum_{k=1}^{n} u_k(x) - f(x) \right| < \varepsilon$$

for all x in X. Show that even this weaker condition[3] is not necessary for continuity of the sum $f(x)$.

Hint. Consider the series in Problem 2.

4. Prove the following proposition, known as the *Heine–Borel theorem*: Any infinite collection S of open intervals covering a closed interval $[a, b]$ has a *finite* subcollection which also covers $[a, b]$.[4]

Hint. Suppose the proposition is false. Then, if $[a, b]$ is divided in half, at least one of the halves, say $[a_1, b_1]$, cannot be covered by a finite number of intervals in S. Dividing $[a_1, b_1]$ in half, we find that at least one of the halves, say $[a_2, b_2]$, cannot be covered by a finite number of intervals in S. Continuing in this way, we get an infinite sequence of intervals $[a_n, b_n]$, none of which can be covered by a finite number of intervals in S. Now use the nested interval theorem[5] to develop a contradiction.

5. Let

$$\sum_{n=1}^{\infty} u_n(x) = u_1(x) + u_2(x) + \cdots + u_n(x) + \cdots \qquad (14)$$

be a functional series converging for all x in a closed interval $X = [a, b]$ to a sum function $f(x)$. Then the series is said to converge *quasi-uniformly* in X to the function $f(x)$ if, given any $\varepsilon > 0$ and any integer $N > 0$, the interval X can be covered by a finite number of open intervals

$$(a_1, b_1), \quad (a_2, b_2), \ldots, \quad (a_m, b_m),$$

with corresponding integers

$$n_1, n_2, \ldots, n_m \, ,$$

---

3. Or, for that matter, the still weaker condition obtained by dropping the stipulation that $n > N$.

4. A set E is said to be *covered* by a collection of sets if E is contained in the union of all the sets in the collection.

5. P. P. Korovkin, *op. cit.*, Theorem 3.5, p. 83.

all greater than N, such that

$$\left| \sum_{k=1}^{n_1} u_k(x) - f(x) \right| < \varepsilon \quad \text{for all } x \text{ in } (a_1, b_1),$$

$$\left| \sum_{k=1}^{n_2} u_k(x) - f(x) \right| < \varepsilon \quad \text{for all } x \text{ in } (a_2, b_2),$$

$$\cdots\cdots\cdots\cdots\cdots\cdots$$

$$\left| \sum_{k=1}^{n_m} u_k(x) - f(x) \right| < \varepsilon \quad \text{for all } x \text{ in } (a_m, b_m).$$

(The condition in Problem 3 corresponds to there being one integer n for all values of x in X, rather than a finite number of different integers $n_1, n_2, \ldots, n_m$, each applying to a suitable set of values of x.) Using this concept, prove the following theorem: Let (14) be a convergent functional series, with sum $f(x)$, each term of which is continuous in an interval $X = [a, b]$. Then a necessary and sufficient condition for $f(x)$ to be continuous in X is that the series converge quasi-uniformly in X.

Hint. To prove the necessity, suppose $f(x)$ is continuous, so that all the remainders $\varphi_n(x)$ are continuous. Let x' be any point of X. Then, given any ε and N, there is an integer $n' > N$ such that $|\varphi_{n'}(x')| < \varepsilon$. Then $|\varphi_{n'}(x)| < \varepsilon$ in some neighborhood $\sigma' = (x' - \delta', x' + \delta')$ of the point x', by the continuity of $\varphi_{n'}(x)$. Let S be the collection of all such open intervals σ', one for each point x' in X. Then S covers the interval X. It follows from the Heine-Borel theorem (Problem 4) that S contains a finite collection $(a_1, b_1), (a_2, b_2), \ldots, (a_m, b_m)$ also covering S, and this is the collection figuring in the definition of the quasi-uniform convergence of (14).

To prove the sufficiency, suppose (14) converges quasi-uniformly to $f(x)$. Given any ε and N, let the intervals (a_1, b_1), $(a_2, b_2), \ldots, (a_m, b_m)$ and the integers $n_1, n_2, \ldots, n_m$ be those figuring in the definition of the quasi-uniform convergence. Given any point x_0 in X, suppose x_0 is contained in the interval

(a_i, b_i). As in the proof of Theorem 1,

$$|f(x) - f(x_0)| \leqslant |f_{n_i}(x) - f_{n_i}(x_0)| + |\varphi_{n_i}(x)| + |\varphi_{n_i}(x_0)|, \qquad (15)$$

where obviously $|\varphi_{n_i}(x_0)| < \varepsilon$. If x also belongs to (a_i, b_i), then $|\varphi_{n_i}(x)| < \varepsilon$ as well. Moreover, we can certainly find a number $\delta > 0$ such that $|x - x_0| < \delta$ implies not only that x belongs to (a_i, b_i) but also that the first term in the right-hand side of (15) is less than ε, so that finally $|f(x) - f(x_0)| < 3\varepsilon$ if $|x - x_0| < \delta$.

6. Consider the series

$$f(x) = \sum_{n=1}^{\infty} \frac{x}{n^p + x^2 n^q}, \qquad (16)$$

where $pq \geqslant 0$ and one of the exponents p and q is greater than 1 (this guarantees the convergence of (16) for all x). Prove that

a) $f(x)$ is continuous in $[0, +\infty)$ if $p > 1$;
b) $f(x)$ is continuous in $(0, +\infty)$ if $p \leqslant 1, q > 1$;
c) $f(x)$ is continuous in $[0, +\infty)$ if $p \leqslant 1, q > 1, p + q > 2$.

Hint. a) Use Weierstrass' test (Theorem 2, p. 14); b) First write (16) in the form

$$\sum_{n=1}^{\infty} \frac{\dfrac{1}{x}}{n^q + \left(\dfrac{1}{x}\right)^2 n^p}.$$

c) Use differentiation to show that the nth term of (16) achieves its maximum, equal to

$$\frac{1}{2} \frac{1}{n^{(p+q)/2}},$$

for $x = n^{(p-q)/2}$.

Comment. It can be shown[6] that $f(x)$ is discontinuous at $x = 0$ in the remaining case, where $p < 1, q > 1, p + q \leqslant 2$.

---

6. See G. M. Fichtenholz, *Improper Integrals* (in The Pocket Mathematical Library), Sec. 14, Example 5.

7. Deduce the logarithmic series

$$\ln (1 + x) = x - \frac{x^2}{2} + \frac{x^3}{3} - \cdots + (-1)^{n-1} \frac{x^n}{n} + \cdots$$

$$(|x| < 1)$$

(*Ruds.*, Example 5, p. 115) from the binomial series

$$(1 + x)^m = 1 + mx + \frac{m(m-1)}{2!} x^2 + \cdots$$

$$+ \frac{m(m-1) \cdots (m - n + 1)}{n!} x^n + \cdots$$

$$(|x| < 1) \tag{17}$$

(*Ruds.*, Sec. 18) by using the familiar formula[7]

$$\ln a = \lim_{k \to \infty} k (\sqrt[k]{a} - 1).$$

Hint. Writing $a = 1 + x$ ($|x| < 1$) and using (17), we find that $\ln (1 + x)$ is the limit as $k \to \infty$ of the expression

$$k [(1 + x)^{1/k} - 1]$$

$$= x - \frac{x^2}{2} \left(1 - \frac{1}{k} \right) + \frac{x^3}{3} \left(1 - \frac{1}{k} \right) \left(1 - \frac{1}{2k} \right) - \cdots$$

$$+ (-1)^{k-1} \frac{x^n}{n} \left(1 - \frac{1}{k} \right) \left(1 - \frac{1}{2k} \right) \cdots \left(1 - \frac{1}{(n-1) k} \right) + \cdots \tag{18}$$

Now apply Theorem 3, after first using Weierstrass' test to establish the uniform convergence of the series (18) for all $k = 1, 2, \ldots$

---

7. P. P. Korovkin, *op. cit.*, p. 102.

8. Deduce the exponential series

$$e^x = 1 + x + \frac{x^2}{2!} + \frac{x^3}{3!} + \cdots + \frac{x^n}{n!} + \cdots$$

(*Ruds.*, Example 1, p. 113) from the binomial series (17) by using the familiar formula[8]

$$e^x = \lim_{k \to \infty} \left(1 + \frac{x}{k}\right)^k.$$

Hint. It follows from (17) that

$$\left(1 + \frac{x}{k}\right)^k = 1 + k\frac{x}{k} + \frac{k(k-1)}{2!}\left(\frac{x}{k}\right)^2 + \cdots$$

$$+ \frac{k(k-1)\cdots(k-n+1)}{k!}\left(\frac{x}{k}\right)^n + \cdots$$

$$= 1 + x + \frac{x^2}{2!}\left(1 - \frac{1}{k}\right)$$

$$+ \frac{x^3}{3!}\left(1 - \frac{1}{k}\right)\left(1 - \frac{2}{k}\right) + \cdots$$

$$+ \frac{x^n}{n!}\left(1 - \frac{1}{k}\right)\left(1 - \frac{2}{k}\right)\cdots\left(1 - \frac{n-1}{k}\right) + \cdots$$

$$(k = 1, 2, \ldots). \qquad (19)$$

Each of these expansions terminates (and in fact contains precisely $k + 1$ terms), but we can regard (19) as an "infinite series" by simply setting all subsequent terms equal to zero. Now apply Theorem 3, after first using Weierstrass' test to establish the uniform convergence of the "series" (19) for all $k = 1, 2, \ldots$

9. Starting from the formula

$$\sin m\theta = m \cos^{m-1}\theta \sin\theta$$

$$- \frac{m(m-1)(m-2)}{3!}\cos^{m-3}\theta \sin^3\theta + \cdots \qquad (20)$$

---

8. P. P. Korovkin, *op. cit.*, p. 69.

(*Ruds.*, p. 128), deduce the power series expansion of $\sin x$:

$$\sin x = x - \frac{x^3}{3!} + \cdots \quad (-\infty < x < \infty) \qquad (21)$$

(*Ruds.*, Example 2, p. 113).

Hint. Setting $\theta = x/m$ in (20), we have

$$\sin x = \cos^m \frac{x}{m} \times$$

$$\times \left[m \tan \frac{x}{m} - \left(1 - \frac{1}{m}\right)\left(1 - \frac{2}{m}\right) \frac{\left(m \tan \dfrac{x}{m}\right)^3}{3!} + \cdots \right].$$
$$(22)$$

As in the preceding problem, the number of terms in the factor in brackets is finite for every m, but increases without limit as $m \to \infty$. Suppose x lies between $-m_0\pi/2$ and $m_0\pi/2$, and let $m > m_0$. Then it is easy to see that $|m \tan (x/m)|$ decreases with increasing m and hence

$$\left| m \tan \frac{x}{m} \right| \leqslant C = m_0 \tan \frac{|x|}{m_0} \quad (m > m_0),$$

so that the factor in brackets in (22) is majorized by the convergent series

$$C + \frac{C^3}{3!} + \cdots .$$

Using Weierstrass' test and Theorem 3, we can now take the limit of the right-hand side of (22). Since $\cos^m (x/m) \to 1$, $m \tan (x/m) \to x$ as $m \to \infty$, this gives (21).

Comment. The power series expansion of $\cos x$ can be found in much the same way. The techniques of Problems 7–9 can be found (in less rigorous form) in the work of Euler (1748).

10. Prove that

a) $\lim\limits_{x \to 1-} \sum\limits_{n=1}^{\infty} \frac{(-1)^{n-1}}{n} \frac{x^n}{1+x^n} = \frac{1}{2} \ln 2;$

b) $\lim\limits_{x \to 1-} (1-x) \sum\limits_{n=1}^{\infty} (-1)^{n-1} \frac{x^n}{1-x^{2n}} = \frac{1}{2} \ln 2.$

Hint. a) Use, Abel's test (Theorem 3, p. 16) to establish the uniform convergence of the series on the left for all x in $(0, 1)$, and then use Theorem 3 to take the limit as $x \to 1-$. In b), after noting that

$$\sum_{n=1}^{\infty} (-1)^{n-1} \frac{(1-x) x^n}{1-x^{2n}}$$

$$= \sum_{n=1}^{\infty} (-1)^{n-1} \frac{x^n}{1+x+x^2+\cdots+x^{2n-1}},$$

use Dirichlet's test (Theorem 4, p. 16) to establish the uniform convergence of the series on the left for all x in $(0, 1)$, and then use Theorem 3 to take the limit as $x \to 1-$.

5. Integration and Differentiation of a Series

Next we consider the problem of integrating the sum of a convergent functional series:

THEOREM 1. *Let*

$$\sum_{n=1}^{\infty} u_n(x) = u_1(x) + u_2(x) + \cdots + u_n(x) + \cdots \qquad (1)$$

be a functional series, each of whose terms $u_n(x)$ is continuous in an interval $X = [a, b]$. Suppose the series (1) converges uniformly in X to a function $f(x)$. Then the series can be integrated *term by term in the sense that*

$$\int_a^b f(x)\, dx = \int_a^b \left\{ \sum_{n=1}^{\infty} u_n(x) \right\} dx = \sum_{n=1}^{\infty} \left\{ \int_a^b u_n(x)\, dx \right\}$$

$$= \int_a^b u_1(x)\, dx + \int_a^b u_2(x)\, dx + \cdots + \int_a^b u_n(x)\, dx + \cdots,$$

$$(2)$$

i.e., the integral of the sum of the series equals the sum of the series made up of the integrals of its terms.

Proof. The existence of all the integrals in (2) follows from the continuity of the functions $u_n(x)$ and $f(x)$ (recall Theorem 1, p. 20). Integrating the identity

$$f(x) = u_1(x) + u_2(x) + \cdots + u_n(x) + \varphi_n(x)$$

in the interval $[a, b]$, we get

$$\int_a^b f(x)\, dx = \int_a^b u_1(x)\, dx + \int_a^b u_2(x)\, dx + \cdots$$

$$+ \int_a^b u_n(x)\, dx + \int_a^b \varphi_n(x)\, dx.$$

Hence the sum of n terms of the series (2) differs from the integral

$$\int_a^b f(x)\, dx$$

by the term

$$\int_a^b \varphi_n(x)\, dx,$$

and to establish the validity of (2), we need only show that

$$\lim_{n \to \infty} \int_a^b \varphi_n(x)\, dx = 0. \qquad (3)$$

Since the series (1) is uniformly convergent, given any $\varepsilon > 0$, there is an integer N such that

$$|\varphi_n(x)| < \varepsilon$$

for all x in X and $n > N$. Hence, for all such n,

$$\left| \int_a^b \varphi_n(x)\, dx \right| \leqslant \int_a^b |\varphi_n(x)|\, dx < \varepsilon\, (b - a),$$

which proves (3). ∎

Remark. The requirement of uniform continuity cannot be dropped in Theorem 1, but in general uniform continuity is only a *sufficient* condition for validity of (2) and not a *necessary* condition (see Problem 1).

Using the theorem on term-by-term integration of series, we can easily prove a related theorem on differentiation of series:

THEOREM 2. *Let* (1) *be a convergent functional series, with sum* $f(x)$, *each of whose terms* $u_n(x)$ *has a continuous derivative* $u_n'(x)$ *in an interval* $X = [a, b]$. *Suppose the series*

$$\sum_{n=1}^{\infty} u_n'(x) = u_1'(x) + u_2'(x) + \cdots + u_n'(x) + \cdots \tag{4}$$

converges uniformly in X *to the function* $\varphi(x)$. *Then the series* (1) *can be differentiated term by term in the sense that* $f(x)$ *is differentiable and*

$$f'(x) = \varphi(x)$$

for all x *in* X, *or equivalently,*

$$\frac{d}{dx} \left\{ \sum_{n=1}^{\infty} u_n(x) \right\} = \sum_{n=1}^{\infty} u_n'(x),$$

i.e., the derivative of the sum of the series equals the sum of the series made up of the derivatives of its terms.

Proof. Using Theorem 1 to integrate (4) term by term from a to x, where x is any point in X, we get

$$\int_a^x \varphi(t)\, dt = \sum_{n=1}^{\infty} \int_a^x u_n'(t)\, dt.$$

Obviously

$$\int_a^x u_n'(t)\, dt = u_n(x) - u_n(a),$$

so that

$$\int_a^x \varphi(t)\, dt = \sum_{n=1}^{\infty} [u_n(x) - u_n(a)] = \sum_{n=1}^{\infty} u_n(x) - \sum_{n=1}^{\infty} u_n(a)$$

$$= f(x) - f(a), \tag{5}$$

where the second equality is justified by the known convergence of the series

$$\sum_{n=1}^{\infty} u_n(x), \qquad \sum_{n=1}^{\infty} u_n(a)$$

(see *Ruds.*, Theorem 4, p. 7). But the integral on the left in (5) is differentiable, with derivative $\varphi(x)$ (*Def. Int.*, Sec. 6, Theorem 2).[9] Hence the function $f(x)$, which differs from this integral by a constant, has the same derivative $\varphi(x)$, i.e., $f'(x) \equiv \varphi(x)$. ∎

Theorem 2 can be freed of certain superfluous assumptions, at the price of a somewhat more complicated proof:

THEOREM 3. *Let* (1) *be a functional series, each of whose terms* $u_n(x)$ *has a finite derivative* $u_n'(x)$ *in an interval* $X = [a, b]$. *Suppose* (1) *converges for at least one value of* x, *say* $x = a$, *while the series of derivatives* (4) *converges uniformly in* X. *Then the series* (1) *converges uniformly in the whole interval* X *to a differentiable function* $f(x)$, *with derivative*

$$f'(x) = \sum_{n=1}^{\infty} u_n'(x).$$

Proof. Choose two distinct points x_0 and x in the interval $[a, b]$, and form the series

$$\sum_{n=1}^{\infty} \frac{u_n(x) - u_n(x_0)}{x - x_0}. \tag{6}$$

---

9. The abbreviation *Def. Int.* refers to the volume *The Definite Integral* by G. M. Fichtenholz (in The Pocket Mathematical Library).

Then, for fixed x_0, the series (6) is uniformly convergent for all $x \neq x_0$ in X. In fact, given any $\varepsilon > 0$, there is a number N such that

$$\left| \sum_{k=n+1}^{n+m} u_n'(x) \right| < \varepsilon \tag{7}$$

for all x in X and all $n > N$, $m = 1, 2, \ldots$, by the uniform convergence of the series (4). Let

$$U(x) = \sum_{k=n+1}^{n+m} u_k(x),$$

where n and m are temporarily fixed. Then, by (7),

$$|U'(x)| < \varepsilon.$$

But

$$\sum_{k=n+1}^{n+m} \frac{u_n(x) - u_n(x_0)}{x - x_0} = \frac{U(x) - U(x_0)}{x - x_0} = U'(c),$$

by the mean value theorem,[10] where $x_0 < c < x$. It follows that

$$\left| \sum_{k=n+1}^{n+m} \frac{u_n(x) - u_n(x_0)}{x - x_0} \right| < \varepsilon$$

for all $x \neq x_0$. Since this inequality holds for all $n > N$ and $m = 1, 2, \ldots$, the series (6) is uniformly convergent for all $x \neq x_0$, as asserted.

The rest of the proof is now straightforward. Choosing $x_0 = a$, we find that the series

$$\sum_{n=1}^{\infty} \frac{u_n(x) - u_n(a)}{x - a}$$

is uniformly convergent, and hence so is the series

$$\sum_{n=1}^{\infty} [u_n(x) - u_n(a)],$$

---

10. See e.g., P.P. Korovkin, *Differentiation* (in The Pocket Mathematical Library), Theorem 2.3, p. 42.

by Corollary 2, p. 14 (with $v(x) = x - a$). But the series

$$\sum_{n=1}^{\infty} u_n(a)$$

is convergent, by hypothesis, and hence the series

$$\sum_{n=1}^{\infty} u_n(x)$$

is uniformly convergent (recall the remark on p. 17). Let $f(x)$ be the sum of this series. Then the sum of the series (6), where x_0 is again any value of x in the interval $X = [a, b]$, is obviously

$$\frac{f(x) - f(x_0)}{x - x_0}.$$

Since (6) is uniformly convergent, we can use Theorem 3, p. 23 to take the limit term by term as $x \to x_0$, obtaining

$$f'(x_0) = \lim_{x \to x_0} \frac{f(x) - f(x_0)}{x - x_0} = \sum_{n=1}^{\infty} \left\{ \lim_{x \to x_0} \frac{u_n(x) - u_n(x_0)}{x - x_0} \right\}$$

$$= \sum_{n=1}^{\infty} u_n'(x_0). \quad \blacksquare$$

Remark. The above theorems on passage to the limit term by term and term-by-term differentiation and integration establish an analogy between functional series and sums of a finite number of functions. However, this analogy is subject to certain provisions, in stating which the notion of *uniform convergence* plays a central role.

PROBLEMS

1. Show that the first of the nonuniformly convergent series

$$\sum_{n=1}^{\infty} \left[\frac{nx}{1 + n^2 x^2} - \frac{(n - 1) x}{1 + (n - 1)^2 x^2} \right]$$

and

$$\sum_{n=1}^{\infty} 2x \left[n^2 e^{-n^2 x^2} - (n-1)^2 e^{-(n-1)^2 x^2} \right],$$

considered in Problem 2, p. 25, can be integrated term by term, while the second cannot.

Hint. In the first case,

$$\lim_{n \to \infty} \int_0^1 \frac{nx}{1 + n^2 x^2}\, dx = \lim_{n \to \infty} \frac{\ln(1 + n^2)}{2n} = 0 = \int_0^1 f(x)\, dx,$$

while in the second case,

$$\lim_{n \to \infty} \int_0^1 2n^2 x e^{-n^2 x^2}\, dx = \lim_{n \to \infty} (1 - e^{-n^2}) = 1 \neq 0 = \int_0^1 f(x)\, dx.$$

2. Prove that the series

$$\frac{1}{1 + x} = 1 - x + x^2 - \cdots + (-1)^n x^n + \cdots \quad (0 \leqslant x < 1)$$

can be integrated term by term in the interval $[0, 1]$, although the series diverges for $x = 1$.

Hint. Clearly,

$$\int_0^1 \frac{dx}{1 + x} = \ln 2 = 1 - \frac{1}{2} + \frac{1}{3} - \cdots + (-1)^n \frac{1}{n} + \cdots$$

(Ruds., p. 117).

3. Prove that Theorem 1 remains true if we replace the word "continuous" by the weaker word "integrable."

Hint. By the uniform convergence of the series, given any $\varepsilon > 0$, there is a value of n so large that

$$|f(x) - f_n(x)| < \frac{\varepsilon}{2},$$

or, equivalently,

$$f_n(x) - \frac{\varepsilon}{2} < f(x) < f_n(x) + \frac{\varepsilon}{2} \tag{8}$$

for all x in $[a, b]$. Partitioning the interval $[a, b]$ into subintervals $[x_{i-1}, x_i]$, where

$$a = x_0 < x_1 < x_2 < \cdots < x_N = b,$$

let l_i be the greatest lower bound and L_i the least upper bound of $f_n(x)$ in $[x_{i-1}, x_i]$, while m_i is the greatest lower bound and M_i the least upper bound of $f(x)$ in $[x_{i-1}, x_i]$, and introduce the "oscillations"

$$\omega_i = L_i - l_i, \quad \Omega_i = M_i - m_i.$$

Because of (8),

$$l_i - \frac{\varepsilon}{2} < f(x) < L_i + \frac{\varepsilon}{2} \quad (i = 1, 2, \ldots, N)$$

for all x in $[x_{i-1}, x_i]$, and hence

$$\Omega_i = M_i - m_i \leqslant L_i - l_i + \varepsilon = \omega_i + \varepsilon.$$

It follows that

$$\sum_{i=1}^{N} \Omega_i \varDelta x_i \leqslant \sum_{i=1}^{N} \omega_i \varDelta x_i + \varepsilon (b - a),$$

where $\varDelta x_i = x_i - x_{i-1}$. But the second term on the right can be made arbitrarily small, while the first term approaches zero as $\lambda = \max \{\varDelta x_1, \varDelta x_2, \ldots, \varDelta x_n\} \to 0$, by the integrability of $f_n(x)$ (*Def. Int.*, Sec. 2, Theorem 4 and subsequent remark). Hence the expression on the left approaches zero as $\lambda \to 0$, thereby implying the integrability of $f(x)$. The rest of the proof is identical with that of Theorem 1.

Comment. The first series in Problem 1 shows that uniform convergence is not a necessary condition for integrability of the sum of a series of integrable functions.

4. Give an example of a nonuniformly convergent series of integrable functions with a nonintegrable sum.

Hint. Let $u_n(x) = 1$ if x is a nonnegative fraction of the form m/n in lowest terms, while $u_n(x) = 0$ if x is any other point of the interval $X = [0, 1]$. Then each $u_n(x)$ is integrable in X, having only a finite number of points of discontinuity (*Def. Int.*, Sec. 3, Theorem 2), but the sum $f(x)$ of the series is the nonintegrable Dirichlet function, equal to 1 if x is rational and 0 if x is irrational (*Def. Int.*, Sec. 1, Prob. 2).

5. Give an example of a series

$$f(x) = \sum_{n=1}^{\infty} u_n(x), \qquad (9)$$

with corresponding series of derivatives

$$\varphi(x) = \sum_{n=1}^{\infty} u_n'(x), \qquad (10)$$

such that $f'(x) \not\equiv \varphi(x)$.

Hint. If

$$f(x) = \sum_{n=1}^{\infty} [e^{-(n-1)^2 x^2} - e^{-n^2 x^2}] \quad (0 \leqslant x \leqslant 1),$$

then we get the nonuniformly convergent differentiated series

$$\varphi(x) = \sum_{n=1}^{\infty} 2x [n e^{-n^2 x^2} - (n-1)^2 e^{-(n-1)^2 x^2}] \quad (0 \leqslant x \leqslant 1)$$

considered in Problem 1, with sum identically equal to zero. Since

$$f(x) = \begin{cases} 0 & \text{if } x = 0, \\ 1 & \text{if } x \neq 0, \end{cases}$$

$f'(0)$ does not exist, and hence we cannot write $f'(x) \equiv \varphi(x)$.

6. Give an example of a series (9) with series of derivatives (10) such that $f'(x) \equiv \varphi(x)$, even though (10) fails to be uniformly convergent.

Hint. If

$$f(x) = \frac{1}{2} \ln (1 + x^2)$$

$$+ \sum_{n=2}^{\infty} \left[\frac{1}{2n} \ln (1 + n^2 x^2) - \frac{1}{2(n-1)} \ln (1 + (n-1)^2 x^2) \right],$$

then we get the nonuniformly convergent differentiated series

$$\varphi(x) = \sum_{n=1}^{\infty} \left[\frac{nx}{1 + n^2x^2} - \frac{(n-1)x}{1 + (n-1)^2x^2} \right] \quad (0 \leqslant x \leqslant 1)$$

considered in Problem 1, with sum identically equal to zero. But $f(x) \equiv 0$ as well, and hence $f'(x) \equiv \varphi(x)$.

Comment. It is clear from Problems 5 and 6 that uniform convergence of the series of derivatives is a sufficient condition for $f'(x) \equiv \varphi(x)$ but not a necessary condition.

7. Let X be an infinite set with a (finite or infinite) limit point a, and let $\{f_n(x)\}$ be a sequence of functions, each of which is defined in X and has a finite limit as x approaches a:

$$\lim_{x \to a} f_n(x) = c_n.$$

Suppose the sequence $\{f_n(x)\}$ converges uniformly in X to a limit function

$$f(x) = \lim_{n \to \infty} f_n(x).$$

Prove that
  1) The limit

$$\lim_{n \to \infty} c_n = c \tag{11}$$

  exists;
  2) The function $f(x)$ approaches c as $x \to a$:

$$\lim_{x \to a} f(x) = c. \tag{12}$$

Hint. This is the sequence analogue of Theorem 3, p. 23.
Comment. It follows from (11) and (12) that

$$\lim_{x \to a} f(x) = \lim_{n \to \infty} c_n,$$

or equivalently,

$$\lim_{x \to a} \lim_{n \to \infty} f_n(x) = \lim_{n \to \infty} \lim_{x \to a} f_n(x).$$

Thus, regarding $f_n(x)$ as a function of *two* variables n and x, we have found conditions under which the order of two limits (one with respect to n, the other with respect to x) can be reversed.

8. State and prove the sequence analogues of Theorems 1 and 2 of Sec. 4.

9. Let $\{f_n(x)\}$ be a sequence of functions, each integrable in an interval $X = [a, b]$. Suppose the sequence $\{f_n(x)\}$ converges uniformly in X to a limit function

$$f(x) = \lim_{n \to \infty} f_n(x).$$

Prove that $f(x)$ is integrable in $[a, b]$, with integral

$$\int_a^b f(x)\, dx = \lim_{n \to \infty} \int_a^b f_n(x)\, dx. \tag{13}$$

Hint. This is the sequence analogue of Problem 3.

Comment. Writing (13) in the form

$$\lim_{n \to \infty} \int_a^b f_n(x)\, dx = \int_a^b \left\{ \lim_{n \to \infty} f_n(x) \right\} dx,$$

we see that under certain conditions the limit on the left can be taken behind the integral sign, in the sense that the limit of the integral of $f_n(x)$ is just the integral of the limit of $f_n(x)$, i.e., we can reverse the operations of taking the limit and integrating. Since integration is itself a limiting operation, this is another instance of the general problem of reversing limiting operations.

10. Let $\{f_n(x)\}$ be a sequence of functions, each differentiable in an interval $X = [a, b]$. Suppose the sequence $\{f_n(x)\}$ converges for at least one value of x, say $x = a$, while the sequence of derivatives $\{f_n(x)\}$ converges uniformly in X. Prove that the sequence $\{f_n(x)\}$ converges uniformly in the whole interval X to a differentiable function $f(x)$, with derivative

$$f'(x) = \lim f_n'(x). \tag{14}$$

Hint. This is the sequence analogue of Theorem 3, p. 35.

Comment. Writing (14) in the form

$$\frac{d}{dx}\left\{\lim_{n\to\infty} f_n(x)\right\} = \lim_{n\to\infty}\left\{\frac{d}{dx} f_n(x)\right\},$$

we see that under certain conditions we can reverse the operations of taking the derivative and taking the limit. Since differentiation is itself a limiting operation, this is still another instance of the general problem of reversing limiting operations.

From the series standpoint, the parameter n in the function $f_n(x)$ cannot, of course, be made more general. From the sequence standpoint, however, we can replace $f_n(x)$ by a function $f(x, y)$ of two variables, where y varies in some infinite set Y with a (finite or infinite) limit point b. The limit as $n \to \infty$ is then replaced by the limit as $y \to b$. The statement and proof of results of this more general kind is not difficult.[11]

6. The Case of Power Series

The most important example of the application of the theory of the preceding sections is the study of the properties of *power series*. We will confine ourselves to power series of the form

$$\sum_{n=0}^{\infty} a_n x^n = a_0 + a_1 x + a_2 x^2 + \cdots + a_n x^n + \cdots, \qquad (1)$$

since series of the more general form

$$\sum_{n=0}^{\infty} a_n (x - x_0)^n = a_0 + a_1 (x - x_0) + a_2 (x - x_0)^2 + \cdots$$
$$+ a_n (x - x_0)^n + \cdots \qquad (2)$$

can be reduced at once to the form (1) by a simple change of variable (cf. *Ruds.*, p. 106).

---

11. See G. M. Fichtenholz, *Parameter-Dependent Integrals* (in The Pocket Mathematical Library).

THEOREM 1. *Suppose the power series* (1) *has radius of convergence* $R > 0$, *and let* r *be any positive number less than* R. *Then the series converges uniformly for all* x *in the closed interval* $[-r, r]$.

Proof. Since $r < R$, the series (1) converges absolutely for $x = r$, i.e., the positive series

$$\sum_{n=0}^{\infty} |a_n| \, r^n = |a_0| + |a_1| \, r + |a_2| \, r^2 + \cdots + |a_n| \, r^n + \cdots \quad (3)$$

is convergent (cf. *Ruds.*, Lemma, p. 68). If $|x| \leqslant r$, the terms of the series (1) do not exceed the corresponding terms of (3) in absolute value, i.e., (3) is a majorant series for (1). Theorem 1 is now an immediate consequence of Weierstrass' test (Theorem 2, p. 14). ■

Remark. Even though the number r can be chosen arbitrarily close to R, Theorem 1 does not imply uniform convergence in the whole interval of convergence $(-R, R)$, as shown by Example, 4, p. 9.

THEOREM 2. *The sum* $f(x)$ *of the power series* (1), *with radius of convergence* R, *is continuous in the interval* $(-R, R)$.

Proof. Given any point $x = x_0$ in $(-R, R)$, we can choose a number $r < R$ such that $|x_0| < r$. By Theorem 1, the series (1) is uniformly convergent in $[-r, r]$. Hence, by Theorem 1, p. 20, the sum $f(x)$ is continuous in $[-r, r]$ and hence, in particular, continuous at $x = x_0$.[12] ■

The continuity of the sum of a power series can be used to prove the following *uniqueness theorem* for power series (analogous to the corresponding theorem for polynomials):

THEOREM 3. *If two power series*

$$\sum_{n=0}^{\infty} a_n x^n = a_0 + a_1 x + a_2 x^2 + \cdots + a_n x^n + \cdots$$

---

12. Note that we have carefully avoided the use of Theorem 1, p. 20 in the whole interval $(-R, R)$, where uniform convergence cannot be guaranteed.

and

$$\sum_{n=0}^{\infty} b_n x^n = b_0 + b_1 x + b_2 x^2 + \cdots + b_n x^n + \cdots$$

have the same sum in a neighborhood of the point $x = 0$,[13] then the series are identical, i.e., coefficients of like powers of x in the two series are equal:

$$a_0 = b_0, \quad a_1 = b_1, \quad a_2 = b_2, \ldots, \quad a_n = b_n, \ldots$$

Proof. Setting $x = 0$ in the identity

$$a_0 + a_1 x + a_2 x^2 + \cdots = b_0 + b_1 x + b_2 x^2 + \cdots, \quad (4)$$

we immediately get $a_0 = b_0$. Dropping the first term in each side of (4) and dividing by x, *under the assumption that $x \neq 0$,* we obtain the new identity

$$a_1 + a_2 x + \cdots = b_1 + b_2 x + \cdots, \quad (5)$$

which holds in a neighborhood of the point $x = 0$ *but not at the point $x = 0$ itself.* Although we are not justified in setting $x = 0$ in (5), we can however let x approach zero. As a result, using the continuity of the sums of the series in (5), implied by Theorem 2, we find that $a_1 = b_1$. Again dropping the first term in each side of (5), dividing by $x \neq 0$ and letting $x \to 0$, we get $a_2 = b_2$, and so on. ∎

Next we consider the more delicate problem of the behavior of a power series near one of the end points $x = \pm R$ of its interval of convergence (henceforth assumed to be finite). We will consider only the right-hand end point $x = R$, since everything said about this case carries over to the case of the left-hand end point by simply changing x to $-x$.

THEOREM 4. *If the power series* (1) *diverges at the end point $x = R$ of its interval of convergence, then the series cannot converge uniformly in the interval* $[0, R)$.

---

13. The theorem holds not only in a *two-sided* neighborhood $(-\delta, \delta)$ of the point $x = 0$, but also in a *one-sided* neighborhood of the form $(-\delta, 0]$ or $[0, \delta)$.

Proof. If the series converged uniformly in $[0, R)$, then, by Theorem 3, p. 23, we could take the limit term by term as $x \to R-$, thereby proving the convergence of the series

$$\sum_{n=0}^{\infty} a_n R^n = a_0 + a_1 R + a_2 R^2 + \cdots + a_n R^n + \cdots,$$

contrary to hypothesis. ∎

We also have the following theorem, which is in effect the converse of Theorem 4:

THEOREM 5. *If the power series* (1) *converges (albeit only conditionally) at the end point* $x = R$ *of its interval of convergence, then the series converges uniformly in the whole interval* $[0, R]$.

Proof. Writing the series (1) in the form

$$\sum_{n=0}^{\infty} a_n x^n = \sum_{n=0}^{\infty} a_n R^n \left(\frac{x}{R}\right)^n \quad (0 \leqslant x \leqslant R),$$

we see that the theorem is an immediate consequence of Abel's test (Theorem 3, p. 6), since the series

$$\sum_{n=0}^{\infty} a_n R^n$$

converges by hypothesis, while the factors $(x/R)^n$ form a non-increasing uniformly bounded sequence:

$$1 \geqslant \frac{x}{R} \geqslant \left(\frac{x}{R}\right)^2 \geqslant \cdots \geqslant \left(\frac{x}{R}\right)^n \geqslant \left(\frac{x}{R}\right)^{n+1} \geqslant \cdots. \quad \blacksquare$$

THEOREM 6 **(Abel's theorem).** *If the power series* (1) *converges or* $x = R$, *then its sum is continuous (from the left) at the point* $x = R$, *i.e.,*

$$\lim_{x \to R-} \sum_{n=0}^{\infty} a_n x^n = \sum_{n=0}^{\infty} a_n R^n.$$

Proof. An immediate consequence of Theorem 5 above and Theorem 1, p. 20. Another proof of Abel's theorem (under the assumption that $R = 1$) has already been given in connection with the Poisson-Abel method of summing divergent series (*Rams.*, Theorem 1, p. 20).[14] ■

We now apply the theorems of Sec. 6 to the case of power series:

THEOREM 7. *The power series*

$$f(x) = \sum_{n=0}^{\infty} a_n x^n, \tag{6}$$

with radius of convergence R, can be integrated term by term in any interval $[0, x]$ *such that* $|x| < R$, *i.e.,*

$$\int_0^x f(x) \, dx = a_0 x + \frac{a_1}{2} x^2 + \frac{a_2}{3} x^3 + \cdots + \frac{a_{n-1}}{n} x^n + \cdots. \tag{7}$$

Moreover, x can coincide with one of the end points of the interval of convergence if the series (1) *converges at the given end point.*

Proof. An immediate consequence of Theorems 1 and 5 above and Theorem 1, p. 32. ■

THEOREM 8. *The power series* (6), *with radius convergence R, can be differentiated term by term inside its interval of convergence, i.e.,*

$$f'(x) = \sum_{n=1}^{\infty} n a_n x^{n-1} = a_1 + 2a_2 x^2 + 3a_3 x^2 + \cdots$$
$$+ n a_n x^{n-1} + \cdots \tag{8}$$

for any x such that $|x| < R$. *This assertion remains true at an end point of the interval of convergence, provided the series* (8) *converges at the given end point.*

Proof. Let x be any point inside the interval of convergence of the series (6), so that $|x| < R$, and let ϱ be any number between $|x|$ and R, so that $|x| < \varrho < R$. Since the series

$$\sum_{n=1}^{\infty} a_n \varrho^n = a_0 + a_1 \varrho + a_2 \varrho^2 + \cdots + a_n \varrho^n + \cdots$$

---

14. The abbreviation *Rams.* refers to the volume *Infinite Series: Ramifications* by G. M. Fichtenholz (in The Pocket Mathematical Library).

converges, its general term is bounded, i.e.,

$$|a_n|\, \varrho^n \leqslant C \quad (n = 1, 2, \ldots)$$

for some constant C. Therefore we get the estimate

$$n\,|a_n|\,|x|^{n-1} = n\,|a_n|\,\varrho^n \left|\frac{x}{\varrho}\right|^{n-1} \frac{1}{\varrho} \leqslant \frac{C}{\varrho}\, n \left|\frac{x}{\varrho}\right|^{n-1}$$

for the absolute value of the nth term of the series (8). But the series

$$\frac{C}{\varrho} \sum_{n=1}^{\infty} n \left|\frac{x}{\varrho}\right|^{n-1} = \frac{C}{\varrho} \left\{ 1 + 2 \left|\frac{x}{\varrho}\right| + \cdots + n \left|\frac{x}{\varrho}\right|^{n-1} + \cdots \right\}$$

converges, as we see at once from D'Alembert's test (*Ruds.*, Theorem 2, p. 24), bearing in mind that $|x/\varrho| < 1$, and hence the series (8) is absolutely convergent. It follows that $R' \geqslant R$, where R' is the radius of convergence of the series (8). Thus given any $r < R$, we have $r < R'$ as well. By Theorem 1, the series (8) converges uniformly in the interval $[-r, r]$, and hence, by Theorem 2, p. 34, the series (6) can be differentiated term by term. Since $r < R$ is arbitrary, this proves the first part of the theorem. As for the second part, suppose the series (8) converges for $x = R$, say. Then, by Theorem 5, the convergence is uniform in the interval $[0, R]$, and Theorem 2, p. 34 can be applied to the whole interval $[0, R]$, i.e., the term-by-term differentiation is permissible for $x = R$ as well. ∎

We have just seen that $R' \geqslant R$. On the other hand, the absolute values of the terms of the original series (6) do not exceed those of the corresponding terms of the series

$$\sum_{n=1}^{\infty} na_n x^n = a_1 x + 2a_2 x^2 + \cdots + na_n x^n + \cdots,$$

which has the same radius of convergence R' as the series (8) It follows that $R \geqslant R'$, and hence finally that $R' = R$. In other words, *the power series (6) and the series (8) obtained from (6) by term-by-term differentiation have the same radius of conver-*

gence. This is also easily proved by using the Cauchy–Hadamard theorem (*Ruds.*, p. 70), since $\sqrt[n]{n} \to 1$ as $n \to \infty$.

THEOREM 9. *The function $f(x)$ represented by a power series in its interval of convergence $(-R, R)$ has derivatives of all orders in $(-R, R)$. Moreover, the power series is just the Taylor series of the function $f(x)$.*

Proof. Clearly, Theorem 8 can be used to differentiate the power series

$$f(x) = a_0 + a_1 x + a_2 x^2 + a_3 x^3 + \cdots + a_n x^n + \cdots$$

repeatedly, since each "differentiated series" is a power series with the same radius of convergence as the original series. This gives

$$f'(x) = a_1 + 2a_2 x + 3a_3 x^2 + \cdots + na_n x^{n-1} + \cdots,$$

$$f''(x) = 1 \cdot 2a_2 + 2 \cdot 3a_3 x + \cdots + (n-1)na_n x^{n-2} + \cdots,$$

$$f'''(x) = 1 \cdot 2 \cdot 3a_3 + \cdots + (n-2)(n-1)na_n x^{n-3} + \cdots,$$

. . . .

$$f^{(n)}(x) = 1 \cdot 2 \cdot 3 \cdots (n-2)(n-1)na_n + \cdots.$$

Setting $x = 0$ in all these formulas, we get the familiar expressions (*Ruds.*, p. 110)

$$a_0 = f(0), \quad a_1 = f'(0), \quad a_2 = \frac{f''(0)}{2!}, \quad a_3 = \frac{f'''(0)}{3!}, \dots,$$

$$a_n = \frac{f^{(n)}(0)}{n!}, \dots \tag{9}$$

for the Taylor coefficients of the function $f(x)$. In the case of the more general power series

$$f(x) = a_0 + a_1 (x - x_0) + a_2 (x - x_0)^2 + \cdots$$
$$+ a_n (x - x_0)^n + \cdots,$$

we need only replace $f(0), f'(0), \dots, f^{(n)}(0), \dots$ by $f(x_0), f'(x_0), \dots, f^{(n)}(x_0), \dots$ in the formulas (9). ∎

Remark. We now see why it was only necessary to consider Taylor series in our earlier study of power series expansions (*Ruds.*, Chap. 5). In fact, if a function has a power series expansion in the first place, then this expansion is necessarily the Taylor series of the sum of the power series. A function $f(x)$ with a Taylor series expansion in powers of $x - x_0$ in some neighborhood of the point x_0 is said to be *analytic* at x_0.

PROBLEMS

1. Prove that the expansion of an even function in a power series

$$\sum_{n=0}^{\infty} a_n x^n = a_0 + a_1 x + a_2 x^2 + \cdots + a_n x^n + \cdots$$

contains only even powers of x, while that of an odd function contains only odd powers of x.

Hint. Use Theorem 3.

2. Suppose it is known that a function $f(x)$ has the expansion

$$f(x) = \sum_{n=0}^{\infty} a_n x^n \quad (-R < x < R),$$

and suppose $f(x)$ is continuous at $x = R$, say, while the series on the right converges for $x = R$. Prove that the expansion continues to hold for $x = R$.

Hint. Use Abel's theorem.

3. Given that

$$\ln (1 + x) = x - \frac{x^2}{2} + \frac{x^3}{3} - \cdots + (-1)^{n-1} \frac{x^n}{n} + \cdots$$

$$(-1 < x < 1)$$

(*Ruds.*, Example 5, p. 115) and that the series

$$1 - \frac{1}{2} + \frac{1}{3} - \cdots + (-1)^{n-1} \frac{1}{n} + \cdots$$

converges (*Ruds.*, Example 1, p. 74), show that

$$\ln 2 = 1 - \frac{1}{2} + \frac{1}{3} - \cdots + (-1)^{n-1} \frac{1}{n} + \cdots.$$

4. Prove that the power series (6) and the "integrated series" (7) have the same radius of convergence.

5. Consider the Dirichlet series

$$\sum_{n=1}^{\infty} \frac{a_n}{n^x}, \qquad (10)$$

with abscissa of convergence $\lambda < \infty$ (*Ruds.*, Example 2, p. 82). Prove that

a) The series (10) converges uniformly for all $x \geq x_0$, where x_0 is any number greater than λ;

b) The sum of (10) is continuous for all $x > \lambda$;

c) If λ is finite and the series

$$\sum_{n=1}^{\infty} \frac{a_n}{n^\lambda} \qquad (11)$$

converges, then (10) converges uniformly for all $x \geq \lambda$;

d) If (11) converges, then the sum of (10) is continuous (from the right) at the point $x = \lambda$.

Comment. These are the analogues for Dirichlet series of Theorems 1, 2, 5 and 6, respectively.

6. Prove that the only continuous solution $f(x) \not\equiv 0$ of the functional equation

$$f(x)f(y) = f(x + y) \qquad (12)$$

in the interval $(-\infty, \infty)$ is of the form

$$f(x) = a^x \quad (a > 0).$$

Hint. If (12) holds, then $f(x)$ is positive for all x, and hence we can take logarithms, obtaining

$$g(x) + g(y) = g(x + y), \qquad (13)$$

where $g(x) = \ln f(x)$. But the only continuous solution $g(x) \neq 0$ of (13) is a linear function of the form

$$g(x) = cx \quad (c = \text{const}).$$

In fact, it follows from (13) that

$$g(rx) = rg(x)$$

for *rational r* and arbitrary x, and hence that

$$g(r) = g(1)r = cr,$$

where $c = g(1)$. Given any irrational ϱ, let $\{r_n\}$ be a sequence of rational numbers converging to ϱ. Then $g(r_n) = cr_n$, and hence

$$\lim_{n \to \infty} g(r_n) = c \lim_{n \to \infty} r_n = c\varrho,$$

or

$$g(\varrho) = c\varrho$$

by the continuity of $g(x)$.

7. Let $E(x)$ be the function defined by the formula

$$E(x) = 1 + \sum_{n=1}^{\infty} \frac{x^n}{n!} \quad (-\infty < x < \infty). \tag{14}$$

Prove that $E(x) = e^x$, without recourse to the Taylor expansion of e^x, given that

$$e = 1 + \sum_{n=1}^{\infty} \frac{1}{n!}.$$

Hint. First prove that

$$E(x)\,E(y) = E(x + y)$$

by multiplying the series (14) by the same series with x replaced by (cf. *Rams.*, Prob. 5, p. 18). Then use Problem 6 and the continuity of $E(x)$.

8. Let $\varphi(m)$ be the function defined by the formula

$$\varphi(m) = 1 + mx + \frac{m(m-1)}{2!} x^2 + \cdots$$

$$+ \frac{m(m-1) \cdots (m-n+1)}{n!} x^n + \cdots \quad (|x| < 1) \quad (15)$$

for arbitrary m and fixed x. Prove that $\varphi(m) = (1+x)^m$, without recourse to the Taylor expansion of $(1+x)^m$.

Hint. First prove that

$$\varphi(m)\, \varphi(m') = \varphi(m + m')$$

by multiplying the series (15) by the same series with m replaced by m'. Then use Problem 6, after proving the continuity of $\varphi(m)$ for all m.

9. In dealing with power series, it has always been tacitly assumed that the terms are arranged in order of increasing powers, although this assumption is unnecessary *inside* the interval of convergence (where the convergence of the series is absolute). Give an example showing the breakdown of Abel's theorem (Theorem 6) if this assumption is not made.

Hint. Consider the series

$$x - \frac{x^2}{2} - \frac{x^4}{4} + \frac{x^3}{3} - \frac{x^6}{6} - \frac{x^8}{8} + \cdots$$

obtained by rearranging the logarithmic series (recall *Rams.*, p. 10).

10. Prove Abel's theorem on the multiplication of series, which states that if the series

$$\sum_{n=1}^{\infty} a_n \qquad (16)$$

and

$$\sum_{n=1}^{\infty} b_n$$

are convergent, with sums A and B, respectively, and if their product

$$\sum_{n=1}^{\infty} c_n$$

(in Cauchy's form) is convergent, with sum C, then $C = AB$.

Hint. It follows from the convergence of (16) that the power series

$$A(x) = \sum_{n=1}^{\infty} a_n x^n$$

is absolutely convergent for $|x| < 1$, since its radius of convergence is certainly no less than 1, and similarly for the series

$$B(x) = \sum_{n=1}^{\infty} b_n x^n,$$

$$C(x) = \sum_{n=1}^{\infty} c_n x^n.$$

Moreover

$$\lim_{x \to 1-} A(x) = A = \sum_{n=1}^{\infty} a_n,$$

by Theorem 2 if $R > 1$ and by Theorem 6 if $R = 1$, and similarly

$$\lim_{x \to 1-} B(x) = B = \sum_{n=1}^{\infty} b_n, \quad \lim_{x \to 1-} C(x) = C = \sum_{n=1}^{\infty} c_n.$$

But

$$A(x) B(x) = C(x) \quad (|x| < 1) \tag{17}$$

by Cauchy's theorem on the multiplication of absolutely convergent series (*Rams.*, p. 14). Taking the limit of both sides of (17) as $x \to 1-$, we get $AB = C$.

Comment. For another proof, see *Rams.*, Theorem 2, p. 29, and for the same proof in a different context, see *Rams.*, p. 111.

11. Prove that a double power series of the form

$$\sum_{j,k=0}^{\infty} a_{jk} (x - x_0)^j (y - y_0)^k$$

(*Rams.*, Sec. 10) can be differentiated term by term any number of times inside its region of convergence. Prove that

$$a_{00} = f(x_0, y_0), \quad a_{10} = \frac{\partial f(x_0, y_0)}{\partial x}, \quad a_{01} = \frac{\partial f(x_0, y_0)}{\partial y},$$

$$a_{20} = \frac{1}{2!} \frac{\partial^2 f(x_0, y_0)}{\partial x^2}, \ldots, \quad a_{jk} = \frac{1}{j! k!} \frac{\partial^{j+k} f(x_0, y_0)}{\partial x^j \partial y^k}, \ldots,$$

where (x_0, y_0) is any point of this region, thereby establishing the *double Taylor series*

$$f(x, y) = \sum_{j,k=0}^{\infty} \frac{1}{j! k!} \frac{\partial^{j+k} f(x_0, y_0)}{\partial x^j \partial y^k} (x - x_0)^j (y - y_0)^k. \qquad (18)$$

Comment. A function $f(x, y)$ with a Taylor series expansion of the form (18) in some neighborhood of the point (x_0, y_0) is said to be *analytic* at (x_0, y_0).

CHAPTER 3

Applications

7. More on Integration of Series

We now give a number of examples further illustrating the technique of integration of series.

Example 1. Find the sum of the series

$$\sum_{n=0}^{\infty} \frac{(-1)^n}{3n + 1}.$$

Solution. Using first Abel's theorem (Theorem 6, p. 46) and then the theorem on term-by-term integration of a power series (Theorem 7, p. 47), we get[1]

$$\sum_{n=0}^{\infty} \frac{(-1)^n}{3n + 1} = \lim_{x \to 1-} \sum_{n=0}^{\infty} \frac{(-1)^n}{3n + 1} x^{3n+1}$$

$$= \lim_{x \to 1-} \int_0^x \sum_{n=0}^{\infty} (-1)^n x^{3n} \, dx$$

$$= \lim_{x \to 1-} \int_0^x \frac{dx}{1 + x^3} = \lim_{x \to 1-} \left\{ \frac{1}{6} \ln \frac{(x + 1)^2}{x^2 - x + 1} \right.$$

$$\left. + \frac{1}{\sqrt{3}} \arctan \frac{2x - 1}{\sqrt{3}} + \frac{\pi}{6\sqrt{3}} \right\}$$

$$= \frac{1}{3} \ln 2 + \frac{\pi}{3\sqrt{3}}.$$

---

1. Concerning the evaluation of the integral, see e.g., G. M. Fichtenholz, *The Indefinite Integral* (in The Pocket Mathematical Library), Chapter 2.

Example 2. By integrating the series

$$\frac{1}{1+x} = 1 - x + x^2 - \cdots + (-1)^{n-1} x^{n-1} + \cdots,$$

$$\frac{1}{1+x^2} = 1 - x^2 + x^4 - \cdots + (-1)^{n-1} x^{2(n-1)} + \cdots$$

in the interval $[0, x]$, where $|x| < 1$, we get the expansions

$$\int_0^x \frac{dx}{1+x} = \ln(1+x)$$

$$= x - \frac{x^2}{2} + \frac{x^3}{3} - \cdots + (-1)^{n-1} \frac{x^n}{n} + \cdots,$$

$$\int_0^x \frac{dx}{1+x^2} = \arctan x$$

$$= x - \frac{x^3}{3} + \frac{x^5}{5} - \cdots + (-1)^{n-1} \frac{x^{2n-1}}{2n-1} + \cdots,$$

which were derived by more complicated means earlier (*Ruds.*, Sec. 17, Examples 4 and 5). The validity of the first expansion for $x = 1$ and of the second expansion for $x = \pm 1$ follows from Abel's theorem.

Example 3. Certain integrals which cannot be expressed in finite form in terms of elementary functions can be expressed as power series by using term-by-term integration. Thus, starting from the expansion

$$e^{-x^2} = 1 - x^2 + \frac{x^4}{2!} - \cdots + (-1)^n \frac{x^{2n}}{n!} + \cdots$$

(*Ruds.*, Example 1, p. 113), we find that

$$\int_0^x e^{-x^2}\, dx = x - \frac{x^3}{3} + \frac{1}{2!}\frac{x^5}{5} - \cdots$$

$$+ (-1)^n \frac{1}{n!}\frac{x^{2n+1}}{2n+1} + \cdots. \tag{1}$$

Similarly, starting from the expansion

$$\frac{\sin x}{x} = 1 - \frac{x^2}{3!} + \frac{x^4}{5!} - \cdots + (-1)^{n-1} \frac{x^{2n-1}}{(2n-1)!} + \cdots$$

(*Ruds.*, Example 2, p. 113), we find that

$$\int_0^x \frac{\sin x}{x}\, dx = x - \frac{x^3}{3!\,3} + \frac{x^5}{5!\,5} - \cdots$$

$$+ (-1)^{n-1} \frac{x^{2n-1}}{(2n-1)!\,(2n-1)} + \cdots. \qquad (2)$$

Example 4. The formulas

$$K(k) = \int_0^{\pi/2} \frac{d\varphi}{\sqrt{1 - k^2 \sin^2 \varphi}} \qquad (0 < k < 1),$$

$$E(k) = \int_0^{\pi/2} \sqrt{1 - k^2 \sin^2 \varphi}\; d\varphi \qquad (0 < k < 1)$$

define the *complete elliptic integrals of the first and second kinds*, respectively, of *modulus k* (cf. *Ruds.*, Prob. 6, p. 93). To expand $K(k)$ in powers of k, we set $x = -k^2 \sin^2 \varphi$ in the expansion

$$\frac{1}{\sqrt{1 + x}} = 1 + \sum_{n=1}^{\infty} (-1)^n \frac{(2n-1)(2n-3)\cdots 3 \cdot 1}{2n(2n-2)\cdots 4 \cdot 2} x^n$$

$$(-1 < x \leqslant 1)$$

(*Ruds.*, Prob. 3c, p. 123). This gives

$$\frac{1}{\sqrt{1 - k^2 \sin^2 \varphi}}$$

$$= 1 + \sum_{n=1}^{\infty} \frac{(2n-1)(2n-3)\cdots 3 \cdot 1}{2n(2n-2)\cdots 4 \cdot 2} k^{2n} \sin^{2n} \varphi, \qquad (3)$$

where the series on the right is uniformly convergent in φ, being majorized for all φ by the convergent series

$$1 + \sum_{n=1}^{\infty} \frac{(2n-1)(2n-3)\cdots 3\cdot 1}{2n(2n-2)\cdots 4\cdot 2} k^{2n}.$$

Hence we can integrate (3) term by term with respect to φ, obtaining

$$K(k) = \frac{\pi}{2}\left\{1 + \sum_{n=1}^{\infty}\left[\frac{(2n-1)(2n-3)\cdots 3\cdot 1}{2n(2n-2)\cdots 4\cdot 2}\right]^2 k^{2n}\right\},$$

after using the formula

$$\int_0^{\pi/2} \sin^{2n}\varphi\, d\varphi = \frac{(2n-1)(2n-3)\cdots 3\cdot 1}{2n(2n-2)\cdots 4\cdot 2}\frac{\pi}{2} \tag{4}$$

(*Def. Int.*, Sec. 9, **Example** 1). Similarly, starting from the expansion

$$\sqrt{1+x} = 1 + \sum_{n=1}^{\infty}(-1)^{n-1}\frac{(2n-3)(2n-5)\cdots 3\cdot 1}{2n(2n-2)\cdots 4\cdot 2} x^n$$

$$(-1 \leqslant x \leqslant 1)$$

(*Ruds.*, Prob. 3b, p. 122), we find that

$$E(k) = \frac{\pi}{2}\left\{1 - \sum_{n=1}^{\infty}\left[\frac{(2n-1)(2n-3)\cdots 3\cdot 1}{2n(2n-2)\cdots 4\cdot 2}\right]^2 \frac{k^{2n}}{2n-1}\right\}. \tag{5}$$

Example 5. If $y = \arcsin(1 - x)$, then

$$\frac{dy}{dx} = -\frac{1}{\sqrt{1-(1-x)^2}} = -\frac{1}{\sqrt{2x}}\frac{1}{\sqrt{1-\dfrac{x}{2}}}$$

$$= -\frac{1}{\sqrt{2x}}\left(1 + \frac{x}{4} + \frac{3}{32}x^2 + \cdots\right)$$

$$= -\frac{1}{\sqrt{2}}x^{-1/2} - \frac{1}{4\sqrt{2}}x^{1/2} - \frac{3}{32\sqrt{2}}x^{3/2} - \cdots$$

(*Ruds.*, Prob. 3c, p. 123). The first term in the series on the right becomes infinite at $x = 0$. Dropping this term, we get a series which is uniformly convergent in any interval $[0, x]$, where $0 < x < 2$ (why?). The indefinite integral of the first term is $-\sqrt{2}\, x^{1/2}$, while that of the rest of the series can be found by integrating term by term. Since we must have $y = \pi/2$ for $x = 0$, we finally get the following expansion in *fractional powers* of x:

$$y = \arcsin x = \frac{\pi}{2} - \sqrt{2}\, x^{1/2} - \frac{1}{6\sqrt{2}}\, x^{3/2} - \frac{3}{80\sqrt{2}}\, x^{5/2} - \cdots$$

$$(0 \leqslant x < 2).$$

Note that y cannot be expanded in positive integral powers of x, since then, by Theorem 8, p. 47, y would have a finite derivative at $x = 0$, which is impossible.

PROBLEMS

1. Deduce the expansion

$$\arcsin x = x + \sum_{n=1}^{\infty} \frac{(2n-1)(2n-3)\cdots 3 \cdot 1}{2n(2n-2)\cdots 4 \cdot 2} \frac{x^{2n+1}}{2n+1}$$

$$(-1 < x < 1) \qquad (6)$$

by term-by-term integration of the expansion

$$\frac{1}{\sqrt{1-x^2}} = 1 + \sum_{n=1}^{\infty} \frac{(2n-1)(2n-3)\cdots 3 \cdot 1}{2n(2n-2)\cdots 4 \cdot 2} x^{2n}$$

$$(-1 < x < 1) \qquad (7)$$

(cf. *Ruds.*, Prob. 3c, p. 123). Prove that (6) also holds for $x = \pm 1$, so that in particular

$$\frac{\pi}{2} = 1 + \sum_{n=1}^{\infty} \frac{(2n-1)(2n-3)\cdots 3 \cdot 1}{2n(2n-2)\cdots 4 \cdot 2} \frac{1}{2n+1}. \qquad (8)$$

Hint. The convergence of the right-hand side of (8) is proved in *Ruds.*, Example 6, p. 30. For an alternative proof, note that

$$x + \sum_{n=1}^{m} \frac{(2n-1)(2n-3)\cdots 3\cdot 1}{2n(2n-2)\cdots 4\cdot 2} \frac{1}{2n+1} < \text{arc sin } x < \frac{\pi}{2}$$

for any *m*. Taking the limit as $x \to 1-$, we get

$$1 + \sum_{n=1}^{m} \frac{(2n-1)(2n-3)\cdots 3\cdot 1}{2n(2n-2)\cdots 4\cdot 2} \frac{1}{2n+1} \leqslant \frac{\pi}{2},$$

from which the convergence follows at once (*Ruds.*, Theorem, p. 9).

2. Deduce the expansion

$$\text{arc sinh } x = x + \sum_{n=1}^{\infty} (-1)^n \frac{(2n-1)(2n-3)\cdots 3\cdot 1}{2n(2n-2)\cdots 4\cdot 2} \frac{x^{2n+1}}{2n+1}$$

$$(-1 \leqslant x \leqslant 1)$$

by term-by-term integration of the expansion

$$\frac{d}{dx} \ln(x + \sqrt{1+x^2}) = \frac{1}{\sqrt{1+x^2}}$$

$$= 1 + \sum_{n=1}^{\infty} (-1)^n \frac{(2n-1)(2n-3)\cdots 3\cdot 1}{2n(2n-2)\cdots 4\cdot 2} x^{2n} \quad (-1<x<1)$$

(*Ruds.*, Prob. 3c, p. 123).

Hint. Note that arc sin $x = \ln(x + \sqrt{1+x^2})$.

3. Use the expansion (1) to calculate the integral

$$I = \int_0^1 e^{-x^2} \, dx$$

to four decimal places.

Hint. Setting $x = 1$ in (1), we get

$$I = 1 - \frac{1}{3} + \frac{1}{10} - \frac{1}{42} + \frac{1}{216} - \frac{1}{1320} + \frac{1}{9360} - \frac{1}{75,600} + \cdots.$$

Thus, if Δ is the error due to dropping all terms from the eighth on, then Δ is negative and

$$|\Delta| < \frac{1}{75,600} < \frac{1.5}{10^5}.$$

Ans. $I = 0.7468\ldots$

4. Use the expansion (2) to calculate the integral

$$I = \int_0^\pi \frac{\sin x}{x}\,dx$$

to within 0.001.

Hint. It is only necessary to retain five terms of the series (2), after setting $x = \pi$.

Ans. $I = 1.852 \pm 0.001$.

5. Represent the following definite integrals as series:

a) $\displaystyle\int_0^1 \frac{\text{arc tan } x}{x}\,dx$;   b) $\displaystyle\int_0^1 x^{-x}\,dx$.

Hint. a) See *Ruds.*, Example 4, p. 114; b) Write x^{-x} as $e^{-x \ln x}$, so that

$$x^{-x} = 1 + \sum_{n=1}^\infty (-1)^n \frac{x^n \ln^n x}{n!}, \qquad (9)$$

where for $x = 0$ we replace the terms of the series starting from $n = 1$ by their limits as $x \to 0+$, i.e., by zeros (by the same token, $x^{-x} = 1$ if $x = 0$). Using differentiation to show that the maximum of the function $|x \ln x|$ is $1/e$, we find that the series (9) is majorized by the convergent series

$$\sum_{n=0}^\infty \frac{(1/e)^n}{n!},$$

so that (9) is uniformly convergent in the interval [0, 1]. Now integrate (9) term by term, using the fact that

$$\int_0^1 x^n \ln^n x \, dx = (-1)^n \frac{n!}{(n+1)^{n+1}}$$

(established by repeated integration by parts).

Ans. a) $1 - \dfrac{1}{3^2} + \dfrac{1}{5^2} - \dfrac{1}{7^2} + \cdots$;

b) $1 + \dfrac{1}{2^2} + \dfrac{1}{3^3} + \dfrac{1}{4^4} + \cdots$.

6. Starting from the expansion

$$\text{arc tan } x = \frac{x}{1+x^2} \sum_{p=0}^{\infty} \frac{2p\,(2p-2)\cdots 4\cdot 2}{(2p+1)(2p-1)\cdots 3\cdot 1} \left(\frac{x^2}{1+x^2}\right)^p$$

$$(0 \leqslant x \leqslant 1)$$

(*Rams.*, Prob. 1, p. 11), prove that

$$\frac{\text{arc sin } y}{\sqrt{1-y^2}} = \sum_{p=0}^{\infty} \frac{2p\,(2p-2)\cdots 4\cdot 2}{(2p+1)(2p-1)\cdots 3\cdot 1} y^{2p+1}$$

$$\left(0 \leqslant y \leqslant \frac{1}{\sqrt{2}}\right). \tag{10}$$

Hint. Set

$$x = \frac{y}{\sqrt{1-y^2}},$$

noting that

$$\text{arc sin } y = \text{arc tan } \frac{y}{\sqrt{1-y^2}}.$$

7. Prove that

a) $\displaystyle\sum_{m=1}^{\infty} \frac{[(m-1)!]^2}{(2m)!} = \frac{\pi^2}{18}$;   b) $\displaystyle\sum_{n=1}^{\infty} \frac{1}{n^2} = \frac{\pi^2}{6}$.

Hint. a) Integrating (10) term by term, we get

$$\frac{1}{2} (\arctan \sin y)^2 = \sum_{p=0}^{\infty} \frac{2p \, (2p-2) \cdots 4 \cdot 2}{(2p+1) \, (2p-1) \cdots 3 \cdot 1} \frac{y^{2p+2}}{2p+2}$$

$$= \sum_{m=1}^{\infty} \frac{(2m-2) \, (2m-4) \cdots 4 \cdot 2}{(2m-1) \, (2m-3) \cdots 3 \cdot 1} \frac{y^{2m}}{2m},$$

or equivalently,

$$2 \, (\arctan \sin y)^2 = \sum_{m=1}^{\infty} \frac{[(m-1)!]^2}{(2m)!} (2y)^{2m},$$

which gives the desired result after setting $y = \frac{1}{2}$; b) See *Rams.*, formula (5), p. 46 or formula (15), p. 85.

8. Evaluate the integral

$$I = \int_0^1 \frac{\ln (1+x)}{x} \, dx.$$

Hint. Expanding $\ln (1+x)$ in power series (*Ruds.*, p. 117), we find that the integrand has the expansion

$$1 - \frac{x}{2} + \frac{x^2}{3} - \cdots + (-1)^{n-1} \frac{x^{n-1}}{n} + \cdots \quad (0 \leqslant x \leqslant 1).$$

Integrating term by term (why is this justified?), we get

$$I = 1 - \frac{1}{2^2} + \frac{1}{3^2} - \cdots + (-1)^{n-1} \frac{1}{n^2} + \cdots$$

$$= \sum_{n=1}^{\infty} (-1)^{n-1} \frac{1}{n^2} = \sum_{n=1}^{\infty} \frac{1}{n^2} - 2 \sum_{n=1}^{\infty} \frac{1}{(2n)^2} = \frac{\pi^2}{12},$$

after using Problem 7b.

9. Evaluate the integral

$$I = \int_0^{\pi} \frac{\ln (1 + a \cos x)}{\cos x} \, dx \quad (|a| < 1)$$

(for $x = \pi/2$ we assign the integrand its limit as $x \to \pi/2$, i.e., the value a).

Hint. Again using the familiar expansion of $\ln (1 + x)$, we have

$$\frac{\ln (1 + a \cos x)}{\cos x} = a + \sum_{n=1}^{\infty} (-1)^n \frac{a^{n+1}}{n+1} \cos^n x \quad (0 \leqslant x \leqslant \pi).$$

Integrating term by term (why is this justified?), we get

$$I = \pi \left\{ a + \sum_{m=1}^{\infty} \frac{(2m-1)(2m-3) \cdots 3 \cdot 1}{2m(2m-2) \cdots 4 \cdot 2} \frac{a^{2m+1}}{2m+1} \right\}, \quad (11)$$

since

$$\int_0^\pi \cos^{2m-1} x \, dx = 0,$$

$$\int_0^\pi \cos^{2m} x \, dx = 2 \int_0^{\pi/2} \cos^{2m} x \, dx$$

$$= \frac{(2m-1)(2m-3) \cdots 3 \cdot 1}{2m(2m-2) \cdots 4 \cdot 2} \pi.$$

Recalling (6), we find that the right-hand side of (11) is just π arc sin a.

10. Prove that

$$\int_{-\pi}^{\pi} \frac{1 - r^2}{1 - 2r \cos x + r^2} \, dx = 2\pi, \quad (12)$$

$$\int_{-\pi}^{\pi} \frac{\cos mx}{1 - 2r \cos x + r^2} \, dx = 2\pi \frac{r^m}{1 - r^2} \quad (|r| < 1). \quad (13)$$

Hint. Start from the expansion

$$\frac{1 - r^2}{1 - 2r \cos x + r^2} = 1 + 2 \sum_{n=1}^{\infty} r^n \cos nx \quad (|r| < 1). \quad (14)$$

5 Fichtenholz (2095)

To prove (14), first multiply both sides by the denominator $1 - 2r \cos x + r^2$, obtaining

$$1 - 2r \cos x + r^2 + 2 \sum_{n=1}^{\infty} r^n \cos nx$$

$$- 2 \sum_{n=1}^{\infty} r^{n+1} 2 \cos nx \cos x + 2 \sum_{n=1}^{\infty} r^{2n+2} \cos nx, \quad (15)$$

and then replace $2 \cos nx \cos x$ by $\cos (n + 1) x + \cos (n - 1)x$, finally reducing (15) to $1 - r^2$. Since the series

$$\sum_{n=1}^{\infty} |r|^n \quad (|r| < 1)$$

converges, the series in the right-hand side of (14) converges uniformly for all x in the interval $[-\pi, \pi]$. Integrating term by term, we get (11), since

$$\int_{-\pi}^{\pi} \cos nx \, dx = 0.$$

Similarly, multiplying both sides of (14) by $\cos mx$ ($m = 1,2,\ldots$) and integrating term by term, we get (13), since

$$\int_{-\pi}^{\pi} \cos mx \cos nx \, dx = \begin{cases} 0 & \text{if} \quad m \neq n, \\ 1 & \text{if} \quad m = n. \end{cases}$$

11. Prove that

$$\int_{0}^{\pi} \ln (1 - 2r \cos x + r^2) \, dx = 0 \quad (|r| < 1). \quad (16)$$

Hint. Writing (14) in the form

$$\frac{\cos x - r}{1 - 2r \cos x + r^2} = \sum_{n=1}^{\infty} r^{n-1} \cos nx, \quad (17)$$

we fix x and regard r as a variable in the interval $(-1, 1)$. Integrating (17) term by term with respect to r, we get

$$\int_0^r \frac{\cos x - r}{1 - 2r \cos x + r^2}\, dr = -\frac{1}{2} \ln (1 - 2r \cos x + r^2)$$

$$= \sum_{n=1}^{\infty} \frac{r^n}{n} \cos nx \quad (|r| < 1),$$

or

$$\ln (1 - 2r \cos x + r^2) = -2 \sum_{n=1}^{\infty} \frac{r^n}{n} \cos nx \quad (|r| < 1). \qquad (18)$$

Fixing r, we can now integrate (18) term by term with respect to x from 0 to π (why is this justified?), obtaining (16).

12. Prove that

$$J_0(x) = \frac{2}{\pi} \int_0^{\pi/2} \cos (x \sin \theta)\, d\theta,$$

$$J_n(x) = \frac{2x^n}{(2n - 1)(2n - 3) \cdots 3 \cdot 1 \cdot \pi} \times$$

$$\times \int_0^{\pi/2} \cos (x \sin \theta) \cos^{2n} \theta\, d\theta \quad (n = 1, 2, \ldots),$$

where $J_n(x)$ is the Bessel function of order n (see *Rams.*, Prob. 13, p. 53).

Hint. For example, integrating the expansion

$$\cos (x \sin \theta) = 1 + \sum_{k=1}^{\infty} (-1)^k \frac{x^{2k} \sin^{2k} \theta}{(2k)!}$$

term by term, with the help of formula (4), we get

$$J_0(x) = 1 + \sum_{k=1}^{\infty} (-1)^k \frac{x^{2k}}{(k!)^2\, 2^{2k}}.$$

13. Use the expansion (5) to calculate $E(1/\sqrt{2})$ to three decimal places.

Hint. We need only retain six terms of the series obtained by setting $k = 1/\sqrt{2}$ in (5).

Ans. $E(1/\sqrt{2}) = 1.350...$

14. Use the expansion (5) to evaluate the integral

$$I = \int_0^{\pi/2} \frac{E(h \sin \theta)}{1 - h^2 \sin^2 \theta} \sin \theta \, d\theta \quad (0 < h < 1).$$

Hint. First prove that

$$\frac{E(k)}{1 - k^2} = \frac{\pi}{2} \left\{ 1 + \left(\frac{1}{2}\right)^2 3k^2 + \left(\frac{1 \cdot 3}{2 \cdot 4}\right)^2 5k^4 \right.$$

$$\left. + \left(\frac{1 \cdot 3 \cdot 5}{2 \cdot 4 \cdot 6}\right)^2 7k^6 + \cdots \right\}$$

$$= \frac{\pi}{2} \left\{ 1 + \sum_{n=1}^{\infty} \left[\frac{(2n-1)(2n-3) \cdots 3 \cdot 1}{2n(2n-2) \cdots 4 \cdot 2}\right]^2 \times \right.$$

$$\left. \times (2n+1) k^{2n} \right\}, \tag{19}$$

by multiplying both sides by $1 - k^2$. Setting $k = h \sin \theta$ in (19) and multiplying by $\sin \theta$, we get a series which converges uniformly for all θ in the interval $[0, \pi/2)$, being majorized by the series (19) with k replaced by h. But

$$\int_0^{\pi/2} \sin^{2n+1} \theta \, d\theta = \frac{2n(2n-2) \cdots 4 \cdot 2}{(2n+1)(2n-1) \cdots 3 \cdot 1}$$

(*Def. Int.*, Sec. 9, Example 1) and hence

$$I = \frac{\pi}{2} \left(1 + \frac{1}{2} h^2 + \frac{1 \cdot 3}{2 \cdot 4} h^4 + \frac{1 \cdot 3 \cdot 5}{2 \cdot 4 \cdot 6} h^6 + \cdots \right)$$

$$= \frac{\pi}{2} \left(1 + \sum_{n=1}^{\infty} \frac{(2n-1)(2n-3) \cdots 3 \cdot 1}{2n(2n-2) \cdots 4 \cdot 2} h^{2n} \right).$$

Comparing this with formula (7), we get

$$I = \frac{\pi}{2} \frac{1}{\sqrt{1 - h^2}} \, .$$

15. Prove that

$$\arc\sin \sqrt{\frac{2x}{1 + x^2}}$$

$$= \sqrt{2}\left(x^{1/2} + \frac{1}{3} x^{3/2} - \frac{1}{5} x^{5/2} - \frac{1}{7} x^{7/2} + \cdots\right)$$

for $0 \leqslant x < 1$.

8. More on Differentiation of Series

Next we give a number of examples illustrating the technique of differentiation of series.

Example 1. Consider the Bessel function of order zero, defined by the power series

$$J_0(x) = 1 + \sum_{k=1}^{\infty} (-1)^k \frac{x^{2k}}{(k!)^2 \, 2^{2k}} \, , \tag{1}$$

as on p. 67. Note that the series in (1) converges for all x (why?). As we now show, this function satisfies the differential equation

$$xy'' + y' + xy = 0, \tag{2}$$

known as *Bessel's equation.* In fact, setting $y = J_0(x)$, we have

$$xy = \sum_{k=1}^{\infty} (-1)^{k-1} \frac{(2k)^2}{(k!)^2 \, 2^{2k}} x^{2k-1},$$

and moreover

$$y' = \sum_{k=1}^{\infty} (-1)^k \frac{2k}{(k!)^2 \, 2^{2k}} x^{2k-1},$$

$$xy'' = \sum_{k=1}^{\infty} (-1)^k \frac{2k \, (2k - 1)}{(k!)^2 \, 2^{2k}} x^{2k-1},$$

after using Theorem 9, p. 49 to differentiate y twice. Adding the last three equations, we find that the coefficient of x^{2k-1} equals

$$\frac{(-1)^k}{(k!)^2\, 2^{2k}} \left[2k\,(2k-1) + 2k - (2k)^2\right] = 0,$$

thereby proving (2).

Example 2. Find the "power series solution" of Bessel's equation (2), i.e., determine the unknown coefficients $a_0, a_1, a_2, \ldots$ of the function

$$y = \sum_{n=0}^{\infty} a_n x^n \quad (-\infty < x < \infty) \tag{3}$$

satisfying (2) for all x.

Solution. Differentiating (3) twice term by term, and substituting the results into equation (2), we find that

$$a_1 + \sum_{n=2}^{\infty} \left[n\,(n-1)\,a_n + n a_n + a_{n-2}\right] x^{n-1} = 0.$$

Hence, by Theorem 3, p. 44,

$$a_1 = 0,$$

$$n^2 a_n + a_{n-2} = 0 \quad (n = 2, 3, \ldots). \tag{4}$$

It follows from the recursion formula (4) that all coefficients with odd indices vanish, i.e.,

$$a_{2k-1} = 0 \quad (k = 1, 2, \ldots) \tag{5}$$

(since a_1 vanishes), while the coefficients with even indices can be expressed in terms of a_0 by the formula

$$a_{2k} = (-1)^k \frac{a_0}{(k!)^2\, 2^{2k}} \quad (k = 0, 1, 2, \ldots). \tag{6}$$

Comparing (3), (5) and (6) with (1), we see that the solution of (2) is just the Bessel function $J_0(x)$, apart from the arbitrary numerical factor a_0. The fact that the series (3) with coefficients

(5) and (6) actually converges for all x can be verified directly, and the function defined by this series obviously satisfies Bessel's equation, by its very construction.

Remark. The technique of the preceding problem (the "method of undetermined coefficients") is familiar from elementary algebra for the case of polynomials (power series with only a finite number of nonzero coefficients). In both the case of polynomials and the more general case of power series, we use the fact that two polynomials or power series are identical if and only if they have the same coefficients.[2]

Example 3. The function $f(x)$ defined by the power series

$$f(x) = \sum_{n=1}^{\infty} \frac{x^n}{n^2} \quad (0 \leqslant x \leqslant 1) \tag{7}$$

satisfies the functional equation

$$f(x) + f(1 - x) + \ln x \ln (1 - x) = C = \text{const}$$

$$(0 < x < 1). \tag{8}$$

To see this, we need only show that the derivative of the left-hand side of (8) vanishes if $f(x)$ is given by (7), i.e., that

$$f'(x) - f'(1 - x) + \frac{1}{x} \ln (1 - x) - \frac{1}{1 - x} \ln x = 0.$$

Differentiating (7) term by term, we get

$$f'(x) = \sum_{n=1}^{\infty} \frac{x^{n-1}}{n} = -\frac{1}{x} \ln (1 - x), \tag{9}$$

since

$$\ln (1 - x) = -x - \frac{x^2}{2} - \frac{x^3}{3} - \cdots,$$

and hence

$$f'(1 - x) = -\frac{1}{1 - x} \ln x, \tag{10}$$

---

2. More exactly, if and only if coefficients of identical powers of x in the two polynomials or power series are equal.

after replacing x by $1 - x$. But together (9) and (10) imply (8). To determine the constant C, we use Abel's theorem (Theorem 6, p. 46) to take the limit of the left-hand side of (8) as $x \to 1-$, obtaining

$$C = \lim_{x \to 1-} f(x) = f(1) = \sum_{n=1}^{\infty} \frac{1}{n^2} = \frac{\pi^2}{6}$$

(see Problem 7b, p. 63).

Example 4. Taking the absolute value of both sides of the infinite product

$$\sin x = x \prod_{n=1}^{\infty} \left(1 - \frac{x^2}{n^2\pi^2} \right)$$

(*Ruds.*, p. 132), we get

$$|\sin x| = |x| \prod_{n=1}^{\infty} \left| 1 - \frac{x^2}{n^2\pi^2} \right|.$$

Provided that $x \neq 0, \pm\pi, \pm 2\pi, \ldots$, we can take logarithms of both sides, obtaining the infinite series

$$\ln |\sin x| = \ln |x| + \sum_{n=1}^{\infty} \ln \left| 1 - \frac{x^2}{n^2\pi^2} \right|.$$

Term-by-term differentiation then gives the expansion

$$\frac{\cos x}{\sin x} = \cot x = \frac{1}{x} + \sum_{n=1}^{\infty} \frac{2x}{x^2 - n^2\pi^2} \qquad (x \neq 0, \pm\pi, \pm 2\pi, \ldots). \tag{11}$$

To justify the differentiation, we need only verify that the series on the right converges uniformly in every (finite) closed interval containing no points of the form $x = 0, \pm\pi, \pm 2\pi, \ldots$ If x belongs to such an interval, then x is bounded, i.e., $|x| < M$, so that, at least for $n > M/\pi$,

$$\left| \frac{2x}{x^2 - n^2\pi^2} \right| = \frac{2|x|}{n^2\pi^2 - |x|^2} < \frac{2M}{n^2\pi^2 - M^2}.$$

But the series

$$\sum_{n > M/\pi} \frac{2M}{n^2\pi^2 - M^2}$$

converges, so that the desired uniform convergence is an immediate consequence of Weierstrass' test (Theorem 2, p. 14).

We can write (11) in the form

$$\frac{\cos x}{\sin x} = \cot x = \frac{1}{x} + \sum_{n=1}^{\infty} \left(\frac{1}{x - n\pi} + \frac{1}{x + n\pi} \right)$$

$$(x \neq 0, \pm\pi, \pm 2\pi, \ldots). \tag{12}$$

This is the expansion of $\cot x$ in *partial fractions*, one fraction for each point at which the denominator $\sin x$ vanishes.

Example 5. Consider the *gamma function*, defined by the formula

$$\Gamma(x) = \frac{1}{x} \prod_{n=1}^{\infty} \frac{\left(1 + \dfrac{1}{n}\right)^x}{1 + \dfrac{x}{n}}, \tag{13}$$

and satisfying the relation

$$\Gamma(x + 1) = x\Gamma(x) \tag{14}$$

(*Ruds.*, p. 102). Recalling that

$$e^C = \prod_{n=1}^{\infty} \frac{e^{1/n}}{1 + \dfrac{1}{n}},$$

where C is Euler's constant (*Ruds.*, Prob. 8, p. 94), we have

$$e^{Cx} = \prod_{n=1}^{\infty} \frac{e^{x/n}}{\left(1 + \dfrac{1}{n}\right)^x}. \tag{15}$$

It follows from (13)–(15) that

$$e^{Cx}\Gamma(x+1) = \prod_{n=1}^{\infty} \frac{e^{x/n}}{1 + \dfrac{x}{n}}, \tag{16}$$

or equivalently,

$$\frac{1}{\Gamma(x+1)} = e^{Cx}\prod_{n=1}^{\infty} \left(1 + \frac{x}{n}\right)e^{-x/n}.$$

Moreover, (14) and (16) imply

$$\ln|\Gamma(x)| = -\ln|x| - Cx + \sum_{n=1}^{\infty}\left(\frac{x}{n} - \ln\left|1 + \frac{x}{n}\right|\right)$$

$$(x \neq 0, -1, -2, \ldots),$$

after taking absolute values and logarithms. Term-by-term differentiation then gives the expansion

$$\frac{\Gamma'(x)}{\Gamma(x)} = -\frac{1}{x} - C + \sum_{n=1}^{\infty}\left(\frac{1}{n} - \frac{1}{x+n}\right)$$

$$(x \neq 0, -1, -2, \ldots). \tag{17}$$

To justify the differentiation, we need only show that the series on the right converges uniformly in every closed interval containing no points of the form $x = 0, -1, -2, \ldots$ If x belongs to such an interval, then x is bounded, i.e., $|x| < M$, so that, at least for $n > M$,

$$\left|\frac{1}{n} - \frac{1}{x+n}\right| = \frac{|x|}{n(n+x)} < \frac{M}{n(n-M)}.$$

But the series

$$\sum_{n>M}\frac{M}{n(n-M)}$$

converges, so that the desired uniform convergence is an immediate consequence of Weierstrass' test. It follows, in particular, that $\Gamma'(x)$ exists except at the points $x = 0, -1, -2, \ldots$

PROBLEMS

1. As in Problem 7, p. 52, let $E(x)$ be the function defined by the formula

$$E(x) = 1 + \sum_{n=1}^{\infty} \frac{x^n}{n!} \quad (-\infty < x < \infty).$$

Use differentiation to prove that $E(x) = e^x$, without recourse to the Taylor expansion of e^x.

Hint. First prove that $E'(x) = E(x)$.

2. Let

$$y = f(x) = 1 + mx + \frac{m(m-1)}{2!} x^2 + \cdots$$

$$+ \frac{m(m-1)\cdots(m-n+1)}{n!} x^n + \cdots, \quad (18)$$

where (unlike Problem 8, p. 53) m is fixed and x varies in the interval $(-1, 1)$. Use differentiation to prove that $f(x) = (1+x)^m$, without recourse to the Taylor expansion of $(1+x)^m$.

Hint. Differentiating (18) term by term, we get

$$y' = f'(x) = m \left\{ 1 + (m-1)x + \frac{(m-1)(m-2)}{2!} + \cdots \right.$$

$$\left. + \frac{(m-1)(m-2)\cdots(m-n)}{n!} x^n + \cdots \right\}.$$

Using the formula

$$\frac{(m-1)(m-2)\cdots(m-n)}{n!} + \frac{(m-1)(m-2)\cdots(m-n+1)}{(n-1)!}$$

$$= \frac{m(m-1)\cdots(m-n+1)}{n!}, \quad (19)$$

we see that

$$(1+x)y' = my,$$

and hence

$$y = C (1 + x)^m,$$

where $C = 1$ since obviously $y = 1$ if $x = 0$.

Comment. The familiar formula

$$C_n^{m-1} + C_{n-1}^{m-1} = C_n^m$$

(cf. *Rams.*, Prob. 1, p. 118) is a special case of (19).

3. Let

$$\varphi(x) = \sum_{n=1}^{\infty} \frac{a_n}{n^x},$$

where the Dirichlet series on the right has abscissa of convergence $\lambda < +\infty$. Then $\varphi(x)$ is continuous for all $x > \lambda$ (recall Problem 5b, p. 51). Prove that

$$\varphi'(x) = -\sum_{n=1}^{\infty} \frac{a_n}{n^x} \ln n \quad (x > \lambda). \tag{20}$$

Hint. Show that the right-hand side of (20) converges uniformly for all $x \geqslant x_0$, where x_0 is any fixed number greater than λ. This can be done by using Abel's test (Theorem 3, p. 16), after writing

$$-\sum_{n=1}^{\infty} \frac{a_n}{n^x} \ln n = -\sum_{n=1}^{\infty} \frac{a_n}{n^{x_0}} \frac{\ln n}{n^{x-x_0}}.$$

Hence, given any $x > \lambda$, there is an interval $[x', x'']$ containing x in which the right-hand side of (20) converges uniformly, so that we can apply Theorem 2, p. 34.

Comment. In the same way, it can be shown that $\varphi(x)$ is infinitely differentiable in the interval $(\lambda, +\infty)$. These considerations apply in particular to the Riemann zeta function

$$\zeta(x) = \sum_{n=1}^{\infty} \frac{1}{n^x},$$

with abscissa of convergence $\lambda = 1$ (*Ruds.*, p. 12).

4. Consider the Bessel function of order n, defined by the power series

$$J_n(x) = \sum_{k=-n}^{\infty} \frac{(-1)^k}{k!\,(k+n)!} \left(\frac{x}{2}\right)^{2k+n},$$

as in *Rams.*, Problem 13, p. 53. Prove that $y = J_n(x)$ satisfies the differential equation

$$x^2 y'' + xy' + (x^2 - n^2)\,y = 0$$

($n = 0, 1, 2, \ldots$), called *Bessel's equation* c.f. p. 69.

5. Show that the *hypergeometric function*, defined by the power series

$$F(\alpha, \beta, \gamma, x)$$
$$= 1 + \sum_{n=1}^{\infty} \frac{\alpha(\alpha+1)\cdots(\alpha+n-1)\beta(\beta+1)\cdots(\beta+n-1)}{n!\,\gamma\,(\gamma+1)\cdots(\gamma+n-1)} \tag{21}$$

(*Ruds.*, pp. 39, 66, 100), satisfies the differential equation

$$x\,(x-1)\,y'' - [\gamma - (\alpha + \beta + 1)\,x]\,y' + \alpha\beta y = 0 \tag{22}$$

known as the *hypergeometric equation*. Conversely, prove that (21) is the solution of (22), to within a numerical factor.

6. Prove that

$$\sum_{n=1}^{\infty} \frac{1}{2^n} \tan \frac{\varphi}{2^n} = \frac{1}{\varphi} - \cot \varphi \quad \left(0 < \varphi < \frac{\pi}{2}\right).$$

Hint. Differentiate the expansion

$$\sum_{n=1}^{\infty} \ln \cos \frac{\varphi}{2^n} = \ln \sin \varphi - \ln \varphi \quad \left(0 < \varphi < \frac{\pi}{2}\right),$$

itself obtained from the easily proved infinite product

$$\prod_{n=1}^{\infty} \cos \frac{\varphi}{2^n} = \frac{\sin \varphi}{\varphi} \quad (\varphi \neq 0)$$

(cf. *Ruds.*, Prob. 3, p. 92).

7. Prove that

a) $\tan x = -\sum_{n=1}^{\infty} \dfrac{2x}{x^2 - \left(n - \dfrac{1}{2}\right)^2 \pi^2}$

$$= -\sum_{n=1}^{\infty} \left[\dfrac{1}{x - \left(n - \dfrac{1}{2}\right)\pi} + \dfrac{1}{x + \left(n - \dfrac{1}{2}\right)\pi}\right];$$

b) $\dfrac{1}{\sin x} = \dfrac{1}{x} + \sum_{n=1}^{\infty} (-1)^n \dfrac{2x}{x^2 - n^2\pi^2}$

$$= \dfrac{1}{x} + \sum_{n=1}^{\infty} (-1)^n \left[\dfrac{1}{x - n\pi} + \dfrac{1}{x + n\pi}\right];$$

c) $\dfrac{1}{\sin^2 x} = \dfrac{1}{x^2} + \sum_{n=1}^{\infty} \left[\dfrac{1}{(x - n\pi)^2} + \dfrac{1}{(x + n\pi)^2}\right]$

(apart from certain exceptional values of x).

Hint. a) Start from $\tan x = -\cot [x - (\pi/2)]$; b) Start from

$$\dfrac{1}{\sin x} = \dfrac{1}{2}\left(\cot \dfrac{x}{2} + \tan \dfrac{x}{2}\right);$$

c) Differentiate (12) term by term.

8. Prove that

a) $\coth x = \dfrac{1}{x} + \sum_{n=1}^{\infty} \dfrac{2x}{x^2 + n^2\pi^2};$

b) $\dfrac{1}{\sinh x} = \dfrac{1}{x} + \sum_{n=1}^{\infty} (-1)^n \dfrac{2x}{x^2 + n^2\pi^2}.$

Hint. Start from *Ruds.*, Problem 2b, p. 133.

9. Prove that

$$\dfrac{\Gamma'(x)}{\Gamma(x)} = -C + \sum_{n=0}^{\infty} \left(\dfrac{1}{n + 1} - \dfrac{1}{x + n}\right) \quad (x \neq 0, -1, -2, \ldots).$$

Hint. Add the series

$$1 + \sum_{n=1}^{\infty} \left(\frac{1}{n+1} - \frac{1}{n} \right) = 0$$

to both sides of (17).

10. Prove that $\Gamma(x)$ is infinitely differentiable (except at the points $x = 0, -1, -2, \ldots$).

9. The Implicit Function Theorem

As another application of the theory of Chapters 1–2, we now prove the following key

THEOREM **(Implicit function theorem).**[3] *Let the function of two variables* $F(x, y)$ *be continuous and have a continuous partial derivative* $F_y(x, y)$ *with respect to* y *in some square* D, *with center* (x_0, y_0), *defined by the inequalities*

$$x_0 - \Delta \leqslant x \leqslant x_0 + \Delta, \quad y_0 - \Delta \leqslant y \leqslant y_0 + \Delta,$$

and suppose that

$$F(x_0, y_0) = 0, \quad F_y(x_0, y_0) \neq 0. \tag{1}$$

Then the equation

$$F(x, y) = 0 \tag{2}$$

defines $y = y(x)$ *in a neighborhood of the point* (x_0, y_0) *as a (single-valued) continuous function of* x *which takes the value* y_0 *for* $x = x_0$.

---

3. This theorem can also be proved by other means, as, for example, in R.A. Silverman, *Modern Calculus and Analytic Geometry,* The Macmillan Co., New York (1969), p. 723. The advantage of the present method of proof (the "method of successive approximations") is that it not only establishes the *existence* of y as a function of x, but also shows how to explicitly *construct* this function. The same technique can be used to prove implicit function theorems of a more general character than the simple case considered here.

Proof. First we consider the special case where (2) takes the form

$$y = y_0 + \varphi(x, y), \tag{2'}$$

where the function of two variables $\varphi(x, y)$ and its partial derivative $\varphi_y(x, y)$ with respect to y are continuous, and the conditions (1) are replaced by

$$\varphi(x_0, y_0) = 0, \quad |\varphi_y(x_0, y_0)| < 1. \tag{1'}$$

Since φ_y is continuous, we can assume from the outset that the domain D is so small that

$$|\varphi_y(x, y)| < \lambda \tag{3}$$

at every point of D, where λ is some constant *less than* 1. Then, without changing the interval in which y varies, we compress the interval in which x varies, replacing it by an interval $[x_0 - \delta, x_0 + \delta]$ so small that the (continuous) function $\varphi(x, y_0)$ of x, which vanishes for $x = x_0$, satisfies the inequality

$$|\varphi(x, y_0)| < (1 - \lambda)\Delta \tag{4}$$

for all x in $[x_0 - \delta, x_0 + \delta]$. This gives us a new square Δ^*, again with center (x_0, y_0), defined by the inequalities

$$x_0 - \delta \leqslant x \leqslant x_0 + \delta, \quad y_0 - \Delta \leqslant y \leqslant y_0 + \Delta,$$

and the considerations that follow all pertain to this square.

Now let y equal the constant y_0 in the right-hand side of (2'), thereby obtaining a new function of x, namely

$$y_1 = y_1(x) = y_0 + \varphi(x, y_0).$$

In the same way, we successively write

$$y_2 = y_2(x) = y_0 + \varphi(x, y_1),$$

$$y_3 = y_3(x) = y_0 + \varphi(x, y_2),$$

$$\cdot \ \cdot \ \cdot \ \cdot \ \cdot \ \cdot \ \cdot \ \cdot \ \cdot \ \cdot \ \cdot \ \cdot$$

$$y_n = y_n(x) = y_0 + \varphi(x, y_{n-1}). \tag{5}$$

As we will soon see, these functions

$$y_1(x), y_2(x), \ldots, y_n(x), \ldots \qquad (6)$$

form a sequence of successive approximations to the desired function $y = y(x)$. We begin by verifying that none of the functions (6) leaves the interval $[y_0 - \Delta, y_0 + \Delta]$, since if any of them were to do so, we could no longer substitute it for y in the right-hand side of (2'). This fact is easily proved by induction. Suppose that

$$y_0 - \Delta \leqslant y_{n-1} \leqslant y_0 + \Delta. \qquad (7)$$

According to (5),

$$y_n - y_0 = \varphi(x, y_{n-1}).$$

But

$$|\varphi(x, y_{n-1})| \leqslant |\varphi(x, y_{n-1}) - \varphi(x, y_0)| + |\varphi(x, y_0)|, \qquad (8)$$

where

$$|\varphi(x, y_{n-1}) - \varphi(x, y_0)| = |\varphi_y(x, \eta)(y_{n-1} - y_0)| < \lambda\Delta \qquad (9)$$

(η between y_0 and y_{n-1}), because of the mean value theorem (with x fixed)[4] and the inequalities (3) and (7). Comparing (4), (8) and (9), we find that

$$|y_n - y_0| = |\varphi(x, y_{n-1})| < \lambda\Delta + (1 - \lambda)\Delta = \Delta,$$

as was to be proved. At the same time, it is easy to see (again by induction) that each of the functions (6) is continuous.

Next we determine the *limit* of the sequence of functions (6). It turns out to be more convenient to consider the series

$$y_0 + \sum_{n=1}^{\infty} (y_n - y_{n-1}), \qquad (10)$$

whose sum is obviously the limit of the sequence (6). From the very definition of (6), we have

$$y_n - y_{n-1} = \varphi(x, y_{n-1}) - \varphi(x, y_{n-2}).$$

---

4. Cf. footnote 10, p. 36.

Again using the mean value theorem and the inequality (3), we find that

$$|y_n - y_{n-1}| < \lambda |y_{n-1} - y_{n-2}|.$$

Replacing n first by $n - 1$, then $n - 2$, and so on, we eventually get

$$|y_n - y_{n-1}| < \lambda^{n-1} |y_1 - y_0| < \lambda^{n-1} (1 - \lambda) \Delta,$$

because of (4). Therefore the series (10) is majorized by the convergent geometric series

$$(1 - \lambda) \Delta \sum_{n=1}^{\infty} \lambda^{n-1}, \tag{11}$$

and hence, by Weierstrass' test (Theorem 2, p. 14), converges *uniformly* for all x in the interval $[x_0 - \delta, x_0 + \delta]$. But then, by Theorem 1, p. 20, the limit function

$$y = y(x) = \lim_{n \to \infty} y_n(x) = y_0(x) + \sum_{n=1}^{\infty} [y_n(x) - y_0(x)]$$

is itself continuous in $[x_0 - \delta, x_0 + \delta]$. Moreover, the fact that $y = y(x)$ satisfies the original equation (2′) is easily verified, by merely using the continuity of $\varphi(x, y)$ to take the limit as $n \to \infty$ in equation (5):

$$y = \lim_{n \to \infty} y_n = y_0 + \lim_{n \to \infty} \varphi(x, y_{n-1}) = y_0 + \varphi(x, y).$$

We must still prove that there are no other values of y except those just found which satisfy (2′), i.e., that (2′) does indeed define $y = y(x)$ as a (single-valued) function of x. Suppose that for some value of x we have

$$\tilde{y} = x_0 + \varphi(x, y),$$

as well as (2′). Then

$$|y - \tilde{y}| = |\varphi(x, y) - \varphi(x, \tilde{y})| < \lambda |y - \tilde{y}|,$$

by the mean value theorem and (3), which is impossible unless $y = \tilde{y}$. In particular,

$$y(x_0) = x_0,$$

as required. This is an immediate consequence of the fact that $y_n(x_0) = x_0$ for all n.

Thus we have proved our theorem for the special case of equation (2'). But the more general case of equation (2) is easily reduced to this special case. In fact, (2) can be written as

$$y = y_0 + \left[y - y_0 - \frac{F(x, y)}{F_y(x_0, y_0)} \right],$$

which is of the form (2') if we set

$$\varphi(x, y) = y - y_0 - \frac{F(x, y)}{F_y(x_0, y_0)}. \tag{12}$$

Note that the function (12) satisfies the conditions (1'), in particular the second condition, since $\varphi_y(x_0, y_0)$ actually vanishes. ∎

PROBLEM

Prove that the error of replacing $y(x)$ by $y_n(x)$ satisfies the estimate

$$|y(x) - y_n(x)| < \Delta \lambda^n.$$

Hint. The remainder after n terms of the series (10) is majorized by the corresponding remainder of the geometric series (11).

10. Analytic Definition of the Trigonometric Functions

Despite the great role that the trigonometric functions play in analysis, they are usually introduced on the basis of purely geometrical considerations, completely foreign to analysis. Therefore a problem of fundamental importance is that of defining the trigonometric functions and studying their basic properties *by purely analytical means*. It turns out that infinite series are precisely the tool enabling us to carry out this program. Thus this section will be devoted to a study of the trigonometric functions, as defined analytically, and will afford still another example of the application of the theory of Chapters 1 and 2.

To this end, we introduce two functions $C(x)$ and $S(x)$, formally defined by the series

$$C(x) = 1 + \sum_{n=1}^{\infty} (-1)^n \frac{x^{2n}}{(2n)!},$$

$$S(x) = \sum_{n=1}^{\infty} (-1)^{n-1} \frac{x^{2n-1}}{(2n-1)!}, \tag{1}$$

which converge for all values of x. For the time being, we make no attempt whatsoever to identify these functions with the familiar functions $\cos x$ and $\sin x$.

As already noted (*Rams.*, Prob. 6, p. 18), the two basic formulas

$$C(x + y) = C(x)\,C(y) - S(x)\,S(y), \tag{2}$$

$$S(x + y) = S(x)\,C(y) + C(x)\,S(y), \tag{3}$$

valid for all x and y, are easily proved by multiplying the series (1) term by term, with appropriate changes of x to y. We now deduce further properties of the functions $C(x)$ and $S(x)$. Replacing x by $-x$, we immediately note that $C(x)$ is an even function, while $S(x)$ is an odd function, i.e.,

$$C(-x) = C(x), \qquad S(-x) = -S(x). \tag{4}$$

Moreover, setting $x = 0$, we find that

$$C(0) = 1, \quad S(0) = 0. \tag{5}$$

Replacing y by $-x$ in (2) and using (4)–(5), we get

$$C^2(x) + S^2(x) = 1. \tag{6}$$

It follows from Theorem 2, p. 44 and Theorem 9, p. 49 that the functions $C(x)$ and $S(x)$ are not only continuous, but also infinitely differentiable. In particular, using Theorem 8, p. 47 to differentiate the series (1) term by term, we easily find that

$$C'(x) = -S(x), \quad S'(x) = C(x). \tag{7}$$

Next we establish the *periodicity* of the functions $C(x)$ and $S(x)$. This will take a bit more work. First we prove that the equation $C(x) = 0$ has a unique root in the interval $(0, 2)$. In fact, $C(0) = 1$ as already noted, while the value of $C(2)$ can be written in the form

$$C(2) = 1 - \frac{2^2}{2!} + \frac{2^4}{4!} - \left(\frac{2^6}{6!} - \frac{2^8}{8!} \right) - \cdots,$$

where we isolate the first three terms of the appropriate series and group the remaining terms in pairs. All the terms in parentheses are positive, since

$$\frac{2^{2n}}{(2n)!} - \frac{2^{2n+2}}{(2n + 2)!} = \frac{2^{2n}}{(2n)!} \left[1 - \frac{2 \cdot 2}{(2n + 1)(2n + 2)} \right] > 0,$$

while the sum of the first three terms is just $-\frac{1}{3}$. Therefore $C(2)$ is certainly negative. But $C(x)$ is continuous, and hence, by the intermediate value theorem,[5] $C(x)$ must vanish at some point of the interval $(0, 2)$. On the other hand, the function

$$S(x) = x \left(1 - \frac{x^2}{2 \cdot 3} \right) + \frac{x^5}{5!} \left(1 - \frac{x^2}{6 \cdot 7} \right) + \cdots$$

is obviously positive in the same interval $(0, 2)$, so that the derivative $C'(x) = -S(x)$ is negative in $(0, 2)$. Therefore the function $C(x)$ decreases steadily as x increases from 0 to 2, and hence can vanish only once in the interval $(0, 2)$. Let $\pi/2$ denote the point at which $C(x)$ vanishes, where the number π is introduced *purely formally* and for the time being is not to be identified with the ratio of the circumference of a circle to its diameter. Thus

$$C\left(\frac{\pi}{2} \right) = 0, \qquad S\left(\frac{\pi}{2} \right) = 1,$$

where the second formula follows from (6), if we bear in mind that $S(x)$ is positive in the interval $(0, \pi)$. Furthermore, setting

---

5. P.P. Korovkin, *Limits and Continuity*, Theorem 3.7, p. 85.

first $x = y = \pi/2$ and then $x = y = \pi$ in formulas (2) and (3), we get

$$C(\pi) = -1, \quad S(\pi) = 0, \quad C(2\pi) = 1, \quad S(2\pi) = 0.$$

Hence, holding x fixed in (2) and (3), we get

$$C(x + \pi) = -C(x), \qquad S(x + \pi) = -S(x) \qquad (8)$$

if $y = \pi$ and

$$C(x + 2\pi) = C(x), \qquad S(x + 2\pi) = S(x)$$

if $y = 2\pi$, i.e., the functions $C(x)$ and $S(x)$ are periodic, with period 2π.

We now show that the functions $C(x)$ and $S(x)$ defined by the series (1) do in fact coincide with the trigonometric functions cos x and sin x, as defined geometrically, at the same time identifying the number π (introduced purely formally) with the familiar number π which plays such an important role in geometry. To this end, we consider the curve specified by the parametric equations

$$x = C(t), \qquad y = S(t) \quad (0 \leqslant t \leqslant 2\pi).$$

Because of (6), every point of this curve satisfies the equation $x^2 + y^2 = 1$, and hence lies on the circle of radius 1 with center at the origin (see Figure 3). Moreover, *every* point P of the circle corresponds to precisely one value of the parameter t in the interval $[0, 2\pi]$, with the exception of the initial point A which corresponds to the two values $t = 0$ and $t = 2\pi$. In fact, as already noted, $S(t) > 0$ for $0 < t \leqslant 2$, and hence, a fortiori, $S(t) > 0$ for $0 < t \leqslant \pi/2$. Changing x to $-t$ in the second of the formulas (8), we get

$$S(\pi - t) = -S(-t) = S(t),$$

from which it follows that $S(t) > 0$ for $\pi/2 \leqslant t < \pi$. This being the case, the function $C(t)$, with derivative $-S(t)$, decreases steadily as t increases from 0 to π, taking each value from 1 to

−1 just once. Hence there is a one-to-one correspondence between the points of the upper half of our circle and values of
the parameter t in the interval $[0, \pi]$. Similarly, because of the
formulas

$$C(t + \pi) = -C(t), \qquad S(t + \pi) = -S(t),$$

obtained by changing x to t in (8), there is a one-to-one correspondence between the points of the lower half of our circle
and values of the parameter t in the interval $[\pi, 2\pi]$.

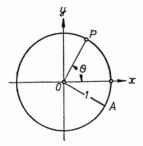

Figure 3

Next we calculate the length of the arc AP of the unit circle,
where P is the point with parameter t. By a familiar formula of
integral calculus,[6]

$$s(t) = \int_0^t \sqrt{[C'(t)]^2 + [S'(t)]^2} \, dt = \int_0^t dt = t, \qquad (9)$$

where we use (6) and (7). It follows from (9) that t is just the
angle $\theta = AOP$, expressed in radians. But then

$$C(\theta) = x = \cos \theta, \qquad S(\theta) = y = \sin \theta,$$

i.e., have finally identified the function $C(\theta)$ with the function
$\cos \theta$ and the function $S(\theta)$ with the function $\sin \theta$. By the same
token, (9) shows that the length of the unit circle is just 2π, so
that our number π can be identified with the number π introduced in elementary geometry.

---

6. See e.g., G.M. Fichtenholz, *Length, Area and Volume* (in The Pocket
Mathematical Library), sec. 4.

PROBLEM

Without identifying $C(x)$ and $S(x)$ with $\cos x$ and $\sin x$, prove that

$$C(2x) = C^2(x) - S^2(x), \qquad S(2x) = 2C(x)\,S(x),$$

$$C\left(\frac{x}{2}\right) = \sqrt{\frac{1 + C(x)}{2}}, \qquad S\left(\frac{x}{2}\right) = \sqrt{\frac{1 - C(x)}{2}},$$

$$C\left(x + \frac{\pi}{2}\right) = -S(x), \qquad S\left(x + \frac{\pi}{2}\right) = C(x).$$

What restrictions must be imposed on the values of x in the formulas for $C(x/2)$ and $S(x/2)$ if the radical denotes the positive square root?

11. A Continuous Nondifferentiable Function

Following van der Waerden, we now construct a function which is continuous for all x, but at the same time nondiffentiable for all x. Let $u_0(x)$ denote the absolute value of the difference between the number x and the nearest integer. Then $u_0(x)$ is *linear* in every interval of the form

$$\left[\frac{s}{2}, \frac{s+1}{2}\right],$$

where s is an integer. Moreover, $u_0(x)$ is continuous and periodic, with period 1. The graph of $u_0(x)$ is the polygonal curve shown in Figure 4a, where each segment of the curve has slope $+1$ or -1.

Next let

$$u_k(x) = \frac{u_0(4^k x)}{4^k} \qquad (k = 1, 2, \ldots).$$

Then $u_k(x)$ is linear in every interval of the form

$$\left[\frac{s}{2 \cdot 4^k} , \frac{s+1}{2 \cdot 4^k} \right],$$

and moreover is continuous and periodic, with period $1/4^k$. The graph of $u_k(x)$ is again a polygonal curve, but with more segments that $u_0(x)$. For example, Figure 4b shows the graph of the function $u_1(x)$, where, as before, every segment of the curve has slope $+1$ or -1.

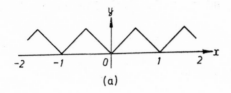

(a)

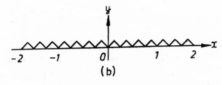

(b)

Figure 4

We now introduce the function

$$f(x) = \sum_{k=0}^{\infty} u_k(x) \quad (-\infty < x < \infty). \tag{1}$$

The series (1) is obviously majorized by the convergent geometric series

$$\sum_{k=0}^{\infty} \frac{1}{2 \cdot 4^k},$$

and hence converges uniformly for all x, by Weierstrass' test (Theorem 2, p. 14). It follows from Theorem 1, p. 20 that $f(x)$ is continuous for all x.

Given any point $x = x_0$, we next observe that

$$\frac{s_n}{2 \cdot 4^n} \leqslant x_0 < \frac{s_n + 1}{2 \cdot 4^n} \quad (n = 0, 1, 2, \ldots)$$

for some integer s_n. Moreover, each closed interval

$$\Delta_n = \left[\frac{s_n}{2 \cdot 4^n}, \frac{s_n + 1}{2 \cdot 4^n} \right]$$

contains the next, and each contains a point x_n whose distance from the point x_0 equals half the length of the interval, i.e.,

$$|x_n - x_0| = \frac{1}{4^{n+1}},$$

where obviously $x_n \to x_0$ as $n \to \infty$. Consider the difference quotient

$$\frac{f(x_n) - f(x_0)}{x_n - x_0} = \sum_{k=0}^{\infty} \frac{u_k(x_n) - u_k(x_0)}{x_n - x_0}.$$

If $k > n$, the number $1/4^{n+1}$ is an integral multiple of the period $1/4^k$ of the function $u_k(x)$, so that $u_k(x_n) = u_k(x_0)$ and the corresponding terms of the series vanish and can be dropped. However, if $k \leqslant n$, the function $u_k(x)$, being linear in the interval Δ_k, is also linear in the interval Δ_n contained in Δ_k, so that

$$\frac{u_k(x_n) - u_k(x_0)}{x_n - x_0} = \pm 1 \quad (k = 0, 1, \ldots, n).$$

Thus, finally,

$$\frac{f(x_n) - f(x_0)}{x_n - x_0} = \sum_{k=0}^{n} (\pm 1),$$

so that the ratio on the left is an even number if n is odd and an odd number if n is even. But then the difference quotient cannot approach a finite limit as $n \to \infty$, i.e., the function (1) cannot have a finite derivative at $x = x_0$ (where x_0 is arbitrary).

PROBLEM

Consider the function

$$f(x) = \sum_{n=0}^{\infty} b^n \cos{(a^n \pi x)},$$

where $0 < b < 1$ and a is an odd number such that

$$ab > 1 + \frac{3\pi}{2}.$$

Show that

a) $f(x)$ is continuous for all x;
b) $f(x)$ is nondifferentiable for all x.

Hint. a) The series is majorized by a convergent geometric series; b) See E. C. Titchmarsh, *Theory of Functions*, second edition, Oxford University Press, New York (1939), Sec. 11.22.

CHAPTER 4

More on Power Series

12. Operations on Power Series

We begin by considering addition, subtraction and multiplication of power series:

THEOREM 1. *Given two power series*

$$\sum_{n=0}^{\infty} a_n x^n = a_0 + a_1 x + a_2 x^2 + \cdots + a_n x^n + \cdots \quad (|x| < r) \quad (1)$$

and

$$\sum_{u=0}^{\infty} b_n x^n = b_0 + b_1 x + b_2 x^2 + \cdots + b_n x^n + \cdots \quad (|x| < r'), \quad (2)$$

with radii of convergence $r > 0$ and $r' > 0$, respectively, let $R = \min \{r, r'\}$.[1] Then

$$\sum_{n=0}^{\infty} a_n x^n \pm \sum_{n=0}^{\infty} b_n x^n = \sum_{n=0}^{\infty} (a_n \pm b_n) x^n \quad (|x| < R) \quad (3)$$

and

$$\sum_{n=0}^{\infty} a_n x^n \sum_{n=0}^{\infty} b_n x^n = \sum_{n=0}^{\infty} (a_0 b_n + a_1 b_{n-1} + \cdots + a_n b_0) x^n$$

$$(|x| < R). \quad (4)$$

Proof. An immediate consequence of the theorem on term-by-term addition or subtraction of series (*Ruds.*, Theorem 4,

---

1. By $\min \{r, r'\}$ is meant the smaller of the two numbers r and r' (or their common value if $r = r'$).

p. 7) and Cauchy's theorem on multiplication of absolutely convergent series (*Rams.*, p. 14). ∎

Example 1. Suppose the series (2) is identical with the series (1). Then it follows from (4) that the series (1) can be *squared* inside its interval of convergence:

$$\left(\sum_{n=0}^{\infty} a_n x^n \right)^2 = \sum_{n=0}^{\infty} (a_0 a_n + a_1 a_{n-1} + \cdots + a_n a_0) x^n \quad (|x| < r). \quad (5)$$

More generally, multiplying (5) by the series (1) again, and repeating this process any number of times, we find that a power series can be raised to any positive power m inside its interval of convergence, and the result is again a power series:

$$\left(\sum_{n=0}^{\infty} a_n x^n \right)^m = \sum_{n=0}^{\infty} a_n^{(m)} x^n \quad (|x| < r, \, m = 1, 2, \ldots). \quad (6)$$

Here the coefficients $a_n^{(m)}$ depend on the coefficients $a_0, a_1, \ldots, a_n$ of the original series, from which they are obtained, as shown by (4), by performing a finite number of operations of addition and multiplication. This observation will be used later, in the proof of the theorem on p. 97.

The next theorem is concerned with the addition of an *infinite* number of power series:

THEOREM 2. *Given infinitely many power series*

$$\sum_{n=0}^{\infty} a_{nm} x^n \quad (m = 0, 1, 2, \ldots),$$

suppose the iterated series

$$\sum_{m=0}^{\infty} \left(\sum_{n=0}^{\infty} a_{nm} x^n \right) \quad (7)$$

is absolutely convergent.[2] *Then the series (7) converges and its sum is the ordinary power series*

$$\sum_{n=0}^{\infty} A_n x^n$$

_____

2. See *Rams.*, Remark, p. 39.

with coefficients

$$A_n = \sum_{m=0}^{\infty} a_{nm} \quad (n = 0, 1, 2, \ldots).$$

Proof. An immediate consequence of the theorem on the equality of the iterated series in the case where one iterated series is absolutely convergent (*Rams.*, Theorem 3, p. 34). ∎

Example 2. Expand the function

$$f(x) = \sum_{m=0}^{\infty} \frac{1}{m!} \frac{a^m}{1 + a^{2m}x^2} \quad (|x| < 1, 0 < a < 1) \quad (8)$$

in powers of x.

Solution. Substituting the series

$$\frac{a^m}{1 + a^{2m}x^2} = \sum_{n=0}^{\infty} (-1)^n a^{(2n+1)m} x^{2n}$$

into (8) and reversing the order of summation, we get

$$f(x) = \sum_{m=0}^{\infty} \frac{1}{m!} \sum_{n=0}^{\infty} (-1)^n a^{(2n+1)m} x^{2n}$$

$$= \sum_{n=0}^{\infty} (-1)^n x^{2n} \sum_{m=0}^{\infty} \frac{a^{(2n+1)m}}{m!} = \sum_{n=0}^{\infty} (-1)^n e^{a^{2n+1}} x^{2n}. \quad (9)$$

According to Theorem 2, this is justified, since the first iterated series is absolutely convergent, as follows from the inequality

$$\sum_{m=0}^{\infty} \frac{1}{m!} \sum_{n=0}^{\infty} a^{(2n+1)m} x^{2n} = \sum_{m=0}^{\infty} \frac{a^m}{m!} \frac{1}{1 - a^{2m}x^2} < \frac{1}{1 - x^2} e^a.$$

Similarly, changing the sign of a in (8) and (9), we find that

$$\sum_{m=0}^{\infty} \frac{(-1)^m}{m!} \frac{a^m}{1 + a^{2m}x^2} = \sum_{m=0}^{\infty} (-1)^n e^{-a^{2n+1}} x^{2n}.$$

Example 3. Changing x to πx in formula (11), p. 72, we get

$$\pi x \cot \pi x = 1 - 2 \sum_{m=1}^{\infty} \frac{x^2}{m^2 - x^2} \quad (x \neq 0, \pm 1, \pm 2, \ldots).$$

If $|x| < 1$, then

$$\frac{x^2}{m^2 - x^2} = \frac{x^2/m^2}{1 - (x^2/m^2)} = \sum_{n=1}^{\infty} \left(\frac{x^2}{m^2}\right)^n \quad (m = 1, 2, \ldots).$$

Since the terms of this series are all positive, it follows at once from Theorem 2 that

$$\sum_{m=1}^{\infty} \frac{x^2}{m^2 - x^2} = \sum_{n=1}^{\infty} s_{2n} x^{2n},$$

where[3]

$$s_n = \sum_{m=1}^{\infty} \frac{1}{m^{2n}} \quad (n = 1, 2, \ldots).$$

Thus the function $\pi x \cot \pi x$ has the power series expansion

$$\pi x \cot \pi x = 1 - 2 \sum_{n=1}^{\infty} s_{2n} x^{2n} \quad (|x| < 1).$$

Example 4. Theorem 2 obviously continues to hold in the case where each of the given (infinitely many) power series reduces to an ordinary polynomial. For example, consider the binomial series

$$(1 + x)^{\alpha} = 1 + \alpha x + \frac{\alpha (\alpha - 1)}{2!} x^2$$

$$+ \frac{\alpha (\alpha - 1) (\alpha - 2)}{3!} x^3 + \cdots \quad (|x| < 1).$$

Fixing x, we can regard the terms of this series as polynomials in α. The series

$$1 + |\alpha| |x| + \frac{|\alpha| (|\alpha| + 1)}{2!} |x|^2$$

$$+ \frac{|\alpha| (|\alpha| + 1) (|\alpha| + 2)}{3!} |x|^3 + \cdots \quad (|x| < 1)$$

---

3. Another expression for the coefficients s_{2n} will be given in Example 3, p. 113.

converges, as follows from D'Alembert's test (*Ruds.*, Theorem 2′, p. 25). Hence, by Theorem 2, we can combine coefficients of like powers of α, obtaining

$$(1 + x)^\alpha = 1 + \alpha \left(x - \frac{x^2}{2} + \frac{x^3}{3} - \cdots \right) + \quad (|x| < 1). \quad (10)$$

On the other hand, we clearly have

$$(1 + x)^\alpha = e^{\alpha \ln (1+x)} = 1 + \alpha \ln (1 + x) + \cdots \quad (11)$$

Comparing (10) and (11), and using Theorem 3, p. 44, we get the familiar expansion

$$\ln (1 + x) = x - \frac{x^2}{2} + \frac{x^3}{3} - \cdots \quad (|x| < 1).$$

PROBLEMS

1. Prove that

$$\pi x \coth \pi x = 1 + 2 \sum_{n=1}^{\infty} (-1)^{n-1} s_{2n} x^{2n},$$

where

$$s_{2n} = \sum_{m=1}^{\infty} \frac{1}{m^{2n}} \quad (n = 1, 2, \ldots).$$

Hint. See Problem 8a, p. 78.

2. Generalize Theorem 2 to the case where the iterated series (7) is of the form

$$\sum_{k,m=0}^{\infty} \left(\sum_{n=0}^{\infty} a_{nkm} x^n \right), \quad (7')$$

i.e., where the "outer series" is a double series.

13. Substitution of One Power Series into Another

As shown by the next theorem, it is often permissible to substitute one power series into another:

THEOREM. *Suppose the function* $y = f(x)$ *has a power series expansion*

$$y = f(x) = \sum_{n=0}^{\infty} a_n x^n$$

$$= a_0 + a_1 x + a_2 x^2 + \cdots + a_n x^n + \cdots, \qquad (1)$$

valid in the interval $(-R, R)$, *while the function* $\varphi(y)$ *has a power series expansion*

$$\varphi(y) = \sum_{m=0}^{\infty} h_m y^m$$

$$= h_0 + h_1 y + h_2 y^2 + \cdots + h_m y^m + \cdots, \qquad (2)$$

valid in the interval $(-\varrho, \varrho)$. *Then, if* $|a_0| = |f(0)| < \varrho$, *the composite function* $\varphi(f(x))$ *is defined in a neighborhood of* $x = 0$, *in which it has the expansion in powers of* x *obtained by replacing* y *in* (2) *by the series* (1), *raising the series* (1) *to the appropriate powers (as in Example 1, p. 93), and finally combining coefficients of like powers of* x.

Proof. If $|a_0| = |f(0)| < \varrho$, then $|f(x)| < \varrho$ for sufficiently small x, and hence it makes sense to talk about the composite function $\varphi(f(x))$. In fact, we then have

$$\varphi(f(x)) = h_0 + \sum_{m=1}^{\infty} h_m \left(\sum_{n=0}^{\infty} a_n x^n \right)^m$$

and hence

$$\varphi(f(x)) = h_0 + \sum_{m=1}^{\infty} \left(\sum_{n=0}^{\infty} h_m a_n^{(m)} x^n \right) \qquad (3)$$

(for sufficiently small $|x|$), where the coefficients $a_n^{(m)}$ are the same as in formula (6), p. 93. If the iterated series in (3) is absolutely convergent, i.e., if the series

$$|h_0| + \sum_{m=1}^{\infty} \left(\sum_{n=0}^{\infty} |h_m|\, |a_n^{(m)}|\, |x|^n \right) \tag{4}$$

is convergent, then we can use Theorem 2, p. 93 to write

$$\varphi(f(x)) = h_0 + \sum_{n=0}^{\infty} \left(\sum_{m=1}^{\infty} h_m a_n^{(m)} \right) x^n$$

for sufficiently small $|x|$, thereby proving the theorem.

To prove the convergence of (4), we first note that the series

$$\sum_{n=0}^{\infty} |a_n|\, |x|^n = |a_0| + |a_1|\, |x| + |a_1|\, |x|^2 + \cdots + |a_n|\, |x|^n + \cdots$$

has a continuous sum (why?), and hence satisfies the inequality

$$\sum_{n=0}^{\infty} |a_n|\, |x|^n < \varrho \tag{5}$$

for sufficiently small $|x|$, since $|a_0| < \varrho$ by hypothesis. Therefore the series

$$|h_0| + \sum_{m=1}^{\infty} |h_m| \left(\sum_{n=0}^{\infty} |a_n|\, |x|^n \right)^m$$

is convergent, with sum A, say. As in formula (6), p. 93, let

$$\left(\sum_{n=0}^{\infty} |a_n|\, |x|^n \right)^m = \sum_{n=0}^{\infty} \alpha_n^{(m)}\, |x|^n,$$

where the coefficients $\alpha_n^{(m)}$ are obtained from $|a_0|, |a_1|, ..., |a_n|$ by performing a finite number of operations of addition and multiplication. Then, clearly,

$$|h_0| + \sum_{m=1}^{\infty} \left(\sum_{n=0}^{\infty} |h_m|\, \alpha_n^{(m)}\, |x|^n \right) = A.$$

But the operations used to obtain $\alpha_n^{(m)}$ from $|a_0|, |a_1|, ..., |a_n|$ are precisely the same as those used to obtain $a_n^{(m)}$ from $a_0, a_1, ..., a_n$, and hence

$$|a_n^{(m)}| \leqslant \alpha_n^{(m)}.$$

It follows that

$$|h_0| + \sum_{m=1}^{\infty} \left(\sum_{n=0}^{\infty} |h_m| \, |a_n^{(m)}| \, |x|^n \right)$$

$$\leqslant |h_0| + \sum_{m=1}^{\infty} \left(\sum_{n=0}^{\infty} |h_m| \, \alpha_n^{(m)} \, |x|^n \right) = A,$$

so that the series (4) is indeed convergent. ∎

Remark 1. Thus the region of values of x in which the above argument guarantees the validity of the expansion of the function $\varphi\,(f(x))$ in powers of x is characterized not only by the obvious condition $|x| < R$, but also by the condition (5). There is no need for the first condition if $R = +\infty$, and no need for the second condition if $\varrho = +\infty$.

Remark 2. In most applications, it is enough to know that the expansion holds for *small* values of $|x|$. A separate investigation can then be made to determine the actual region of applicability of the expansion.

Example 1. To illustrate Remark 2, let

$$\varphi(y) = \sum_{m=0}^{\infty} y^m$$

(so that $\varrho = 1$), and substitute the function

$$f(x) = 2x - x^2$$

for y (here, of course, $R = +\infty$). Then it makes sense to write

$$\varphi\,(f(x)) = \frac{1}{1 - (2x - x^2)} = \frac{1}{(1 - x)^2},$$

provided only that

$$-1 < 2x - x^2 < 1,$$

or equivalently,

$$1 - \sqrt{2} < x < 1 + \sqrt{2} \quad (x \neq 1). \tag{6}$$

The expansion of this function in powers of x is just

$$\frac{1}{(1 - x)^2} = 1 + 2x + 3x^2 + 4x^3 + \cdots, \tag{7}$$

as can be found by multiplying the series

$$\frac{1}{1 - x} = 1 + x + x^2 + x^3 + \cdots \tag{8}$$

by itself, or alternatively, by using Theorem 8, p. 43 to differentiate (8) term by term. The series (7) converges for $-1 < x < 1$. Comparing this fact with (6), we see that the expansion

$$\sum_{m=0}^{\infty} (2x - x^2)^m = 1 + 2x + 3x^2 + \cdots \tag{9}$$

is valid if

$$1 - \sqrt{2} < x < 1. \tag{10}$$

It is interesting to compare (10) with the region of applicability of (9) implied by the condition (5), as in Remark 1, which leads to the requirement

$$2|x| + |x|^2 < 1$$

or

$$1 - \sqrt{2} < x < \sqrt{2} - 1.$$

As we see, the expansion (9) actually holds in the larger interval (10).

Example 2. Deduce the binomial series from the logarithmic and exponential series.

Solution. Clearly

$$(1 + x)^\alpha = e^{\alpha \ln (1+x)} = \exp \left\{ \alpha \left(x - \frac{x^2}{2} + \frac{x^3}{3} - \cdots \right) \right\}$$

$$= 1 + \alpha \left(x - \frac{x^2}{2} + \frac{x^3}{3} - \cdots \right)$$

$$+ \frac{\alpha^2}{2!} \left(x - \frac{x^2}{2} + \frac{x^3}{3} - \cdots \right)^2 + \cdots$$

$$= 1 + \alpha x + \frac{\alpha (\alpha - 1)}{2!} x^2 + \cdots \quad (\exp x \equiv e^x)$$

$$\tag{11}$$

for $|x| < 1$ and arbitrary α. Thus the first few terms of the expansion of $(1 + x)^\alpha$ in powers of x are easily found. To find the coefficient of x^n, we argue as follows: Clearly this coefficient is some polynomial $Q_n(\alpha)$ of degree n in α. Since there is no term involving x^n in the expansion (11) if $\alpha = 0, 1, \ldots, n - 1$, the polynomial must vanish at the indicated points. But then, by the familiar factorization theorem for polynomials,

$$Q_n(\alpha) = c\alpha\,(\alpha - 1) \cdots (\alpha - n - 1),$$

where c is a constant. For $\alpha = n$, the coefficient of x^n is just 1, and hence $Q_n(n) = 1$. It follows that

$$c = \frac{1}{n!},$$

so that finally

$$Q_n(\alpha) = \frac{\alpha\,(\alpha - 1) \cdots (\alpha - n - 1)}{n!},$$

as in the binomial series.

Example 3. Let $f(x)$ be a function with an expansion

$$f(x) = a_1 x + a_2 x^2 + a_3 x^3 + \cdots + a_n x^n + \cdots \tag{12}$$

in powers of x, without a constant term. Then, by the theorem on substitution of one power series into another, the function $g(x) = e^{f(x)}$ has an expansion

$$g(x) = e^{f(x)} = 1 + b_1 x + b_2 x^2 + b_3 x^3 + \cdots$$
$$+ b_n x^n + \cdots \tag{13}$$

for the same values of x, with constant term 1 since $g(0) = e^{f(0)} = e^0 = 1$. To find this expansion, we use the method of undetermined coefficients (see the remark on p. 71) to express the unknown coefficients $b_1, b_2, \ldots$ in terms of the known coefficients $a_1, a_2, \ldots$. First we differentiate (12) and (13), obtaining

$$f'(x) = a_1 + 2a_2 x + 3a_3 x^2 + \cdots + na_n x^{n-1} + \cdots, \tag{14}$$

$$e^{f(x)} f'(x) = b_1 + 2b_2 x + 3b_3 x^2 + \cdots + nb_n x^{n-1} + \cdots. \tag{15}$$

Then, replacing the factors in the left-hand side of (15) by their expansions (13) and (14), we find that

$$(1 + b_1 x + b_2 x^2 + b_3 x^3 + \cdots)(a_1 + 2a_2 x + 3a_3 x^2 + \cdots)$$

$$= b_1 + 2b_2 x + 3b_3 x^2 + \cdots.$$

Equating coefficients of like powers of x, we arrive at the system of equations

$$a_1 = b_1, \quad 2a_2 + a_1 b_1 = 2b_2, \quad 3a_3 + 2a_2 b_1 + a_1 b_2 = 3b_3, \ldots,$$

$$na_n + (n - 1) a_{n-1} b_1 + \cdots + 2a_2 b_{n-2} + a_1 b_{n-1} = nb_n, \ldots,$$

$$(16)$$

from which the unknown coefficients $b_1, b_2, \ldots$ can be determined *recursively*.

Example 4. Prove that the expansion of the function

$$g(x) = (1 - x) \exp \left\{ x + \frac{x^2}{2} + \cdots + \frac{x^{m-1}}{m - 1} \right\}$$

in powers of x begins like

$$1 - \frac{x^m}{m} + \cdots,$$

and that all the coefficients of the expansion are less than 1 in absolute value.

Solution. First we note that $g(x)$ can be written in the form $g(x) = e^{f(x)}$, where

$$f(x) = \ln (1 - x) + x + \frac{x^2}{2} + \cdots + \frac{x^{m-1}}{m - 1}$$

$$= -\frac{x^m}{m} - \frac{x^{m+1}}{m + 1} - \cdots.$$

Thus $f(0) = 0$, as in Example 3, so that $f(x)$ has an expansion of the form (12) and $g(x)$ has an expansion of the form (13). The first assertion is an immediate consequence of the formula

$$g(x) = \exp \left\{ -\frac{x^m}{m} - \frac{x^{m+1}}{m + 1} - \cdots \right\},$$

while the second assertion can be proved *inductively*. In fact, suppose every coefficient b_i with index less than n in the expansion (13) has absolute value less than 1. Since in the present case

$$a_k = \begin{cases} 0 & \text{if } k < m, \\[2ex] -\dfrac{1}{k} & \text{if } k \geqslant m, \end{cases}$$

it then follows from the nth of the equations (16) that

$$|b_n| \leqslant |a_n| + \frac{n-1}{n}|a_{n-1}| + \cdots + \frac{2}{n}|a_2| + \frac{1}{n}|a_1|$$

$$= \frac{1}{n} + \frac{n-1}{n}\frac{1}{n-1} + \cdots + \frac{m}{n}\frac{1}{n} < 1.$$

Example 5. Show that the infinite product

$$F(x) = \prod_{m=1}^{\infty} (1 + q^m x) = (1 + qx)(1 + q^2 x)(1 + q^3 x) \cdots$$
$$(|q| < 1)$$

has an expansion in powers of x for sufficiently small $|x|$, and find the coefficients of this expansion.

Solution. The infinite product converges and has a positive value if $|x| < 1$ (see *Ruds.*, Theorem 5, p. 96). Taking logarithms, we get

$$\ln F(x) = \sum_{m=1}^{\infty} \ln(1 + q^n x) = \sum_{m=1}^{\infty} \left(q^m x - \frac{1}{2} q^{2m} x^2 + \cdots \right).$$

In particular, the series on the right continues to converge if we replace all the terms in parentheses by their absolute values (why?). It follows from Theorem 2, p. 93 that $\ln F(x)$ has an expansion in powers of x in a neighborhood of zero, and hence, by the theorem on p. 97, the same is true of the function $F(x) = e^{\ln F(x)}$. Thus, for sufficiently small $|x|$, we have

$$F(x) = 1 + b_1 x + b_2 x^2 + \cdots + b_n x^n + \cdots, \qquad (17)$$

as in Example 3, where the coefficients $b_1, b_2, ..., b_n, ...$ are still to be determined. The simplest way of doing this is to start from the obvious identity

$$F(x) = (1 + qx) F(qx),$$

which takes the form

$$1 + b_1 x + b_2 x^2 + \cdots + b_n x^n + \cdots$$

$$= (1 + qx)(1 + b_1 qx + b_2 q^2 x^2 + \cdots + b_n q^n x^n + \cdots)$$

after using (17). Using Theorem 3, p. 44 to equate coefficients of like powers of x, we have

$$b_1 q + q = b_1, \qquad b_2 q^2 + b_1 q^2 = b_2, ...,$$

$$b_n q^n + b_{n-1} q^n = b_n, ...,$$

or

$$b_1 = \frac{q}{1-q}, \qquad b_2 = \frac{b_1 q^2}{1-q^2}, ..., \qquad b_n = \frac{b_{n-1} q^n}{1-q^2}, ...,$$

so that finally

$$b_1 = \frac{q}{1-q}, \qquad b_2 = \frac{q^3}{(1-q)(1-q^2)}, ...,$$

$$b_n = \frac{q^{n(n-1)/2}}{(1-q)(1-q^2)\cdots(1-q^n)}, \cdots$$

Example 6. Expand the function

$$\frac{1}{\sqrt{1 - 2x\alpha + \alpha^2}} = [1 + (\alpha^2 - 2x\alpha)]^{-1/2} \qquad (18)$$

in powers of α, for arbitrary fixed x.

Solution. The possibility of such an expansion is guaranteed by the theorem on p. 97 if

$$|\alpha|^2 + 2 |x| |\alpha| < 1.$$

It is easy to see that the coefficient of α^n is some polynomial $P_n = P_n(x)$ of degree n, so that

$$\frac{1}{\sqrt{1 - 2x\alpha + \alpha^2}} = 1 + P_1\alpha + P_2\alpha^2 + \cdots + P_n\alpha^n + \cdots.$$

$$(19)$$

To determine these coefficients, we differentiate (19) with respect to α, obtaining

$$\frac{x - \alpha}{\sqrt{(1 - 2x\alpha + \alpha^2)^3}} = P_1 + 2P_2\alpha + \cdots + nP_n\alpha^{n-1} + \cdots.$$

$$(20)$$

Comparing (19) and (20), we see that

$$(1 - 2x\alpha + \alpha^2)(P_1 + 2P_2\alpha + \cdots + nP_n\alpha^{n-1} + \cdots)$$

$$= (x - \alpha)(1 + P_1\alpha + P_2\alpha^2 + \cdots + P_n\alpha^n + \cdots).$$

Equating coefficients of like powers of α, we get $P_1 = x$,

$$2P_2 x - 2xP_1 = -1 + xP_1,$$

i.e.,

$$P_2 = \frac{3x^2 - 1}{2},$$

and, in general,

$$(n + 1)P_{n+1} - 2nxP_n + (n - 1)P_{n-1} = xP_n - P_{n-1},$$

or

$$(n + 1)P_{n+1} - (2n + 1)xP_n + nP_{n-1} = 0.$$

$$(21)$$

Having just found the first two polynomials P_1 and P_2, we can use the recursion formula (21) to find the remaining polynomials. Thus, for example, setting $n = 2, 3$ in (21), we get

$$P_3 = \frac{5x^3 - 3x}{2}, \qquad P_4 = \frac{35x^4 - 30x^2 + 3}{8}.$$

The polynomials $P_n(x)$ are known as the *Legendre polynomials* (cf. *Def. Int.*, Sec. 13, in particular Formula (10)). Correspondingly, the function of two variables (18) is called the *generating function* of the Legendre polynomials. Various properties of the Legendre polynomials can be deduced from the expansion (19).

PROBLEMS

1. Prove that

$$\frac{1}{e}(1+x)^{1/x} = 1 - \frac{1}{2}x + \frac{11}{24}x^2 - \frac{7}{16}x^3 + \frac{2447}{5760}x^4$$
$$- \frac{959}{2304}x^5 + \cdots$$

for $|x| < 1$.

2. Apply the method of Example 3 to Examples 1 and 2.

3. Given an expansion of the form

$$g(x) = 1 + b_1 x + b_2 x^2 + b_3 x^3 + \cdots + b_n x^n + \cdots,$$

find the coefficients of the expansion

$$f(x) = \ln g(x) = a_1 x + a_2 x^2 + a_3 x^3 + \cdots + a_n x^n + \cdots.$$

Ans. The coefficients $a_1, a_2, \ldots$ and $b_1, b_2, \ldots$ satisfy the same formulas (16) as before, but this time it is the coefficients $a_1, a_2, \ldots$ that are unknown and the coefficients $b_1, b_2, \ldots$ that are known, rather than vice versa.

4. Prove that

$$\sum_{n=1}^{\infty}\left(\frac{x^2}{n^2\pi^2} + \frac{1}{2}\frac{x^4}{n^4\pi^4} + \cdots\right) = \left(\frac{x^2}{6} - \frac{x^4}{120} + \cdots\right)$$
$$+ \frac{1}{2}\left(\frac{x^2}{6} - \frac{x^4}{120} + \cdots\right)^2 + \cdots, \tag{22}$$

and hence that

$$\sum_{n=1}^{\infty} \frac{1}{n^2} = \frac{\pi^2}{6}, \qquad \sum_{n=1}^{\infty} \frac{1}{n^4} = \frac{\pi^4}{90}, \cdots \qquad (23)$$

Hint. Noting that

$$\frac{\sin x}{x} = \prod_{n=1}^{\infty} \left(1 - \frac{x^2}{n^2\pi^2}\right) = 1 - \frac{x^2}{6} + \frac{x^4}{120} - \cdots$$

(*Ruds.*, pp. 114, 132), take logarithms (*Ruds.*, Theorem 4, p. 95). This gives

$$\sum_{n=1}^{\infty} \ln\left(1 - \frac{x^2}{n^2\pi^2}\right) = \ln\left(1 - \frac{x^2}{6} + \frac{x^4}{120} - \cdots\right),$$

which implies (23). Now compare coefficients of like powers of x in (22).

Comment. We will prove (23) by another method in Example 3, p. 113.

5. Prove that if the function $f(x)$ has an expansion

$$f(x) = \sum_{n=0}^{\infty} a_n x^n = a_0 + a_1 x + a_2 x^2 + \cdots + a_n x^n + \cdots \qquad (24)$$

in powers of x valid in an interval $(-R, R)$ and if $\bar{x}$ is any point of $(-R, R)$, then $f(x)$ has an expansion in powers of $x - \bar{x}$ in some neighborhood of $\bar{x}$.

Hint. Setting $x = \bar{x} + y$, we can use the theorem on p. 97 (with the roles of x and y reversed) to go over to an expansion

$$\sum_{k=0}^{\infty} A_k y^k = \sum_{k=0}^{\infty} A_k (x - \bar{x})^k$$

in powers of y, provided that $|\bar{x}| + |y| < R$, i.e., $|y| < R - |\bar{x}|$. Carrying out the operations called for in the series

$$\sum_{n=1}^{\infty} a_n (\bar{x} + y)^n,$$

we can easily find the coefficients $A_0, A_1, \ldots, A_n, \ldots$. In fact,

$$A_0 = \sum_{n=0}^{\infty} a_n \bar{x}^n = f(\bar{x}),$$

and in general,

$$A_k = \sum_{n=k}^{\infty} \frac{n(n-1)\cdots(n-k+1)}{k!} a_n \bar{x}^{n-k}$$

$$= \frac{1}{k!} \sum_{n=k}^{\infty} n(n-1)\cdots(n-k+1) a_n \bar{x}^{n-k} = \frac{f^{(k)}(\bar{x})}{k!},$$

a result which is hardly surprising, in view of Theorem 9, p. 49.

   Comment. Here we have chosen a series (24) in powers of x only for simplicity, and our considerations have an obvious analogue for the case of a series in powers of $x - x_0$. It will be recalled from the remark on p. 50 that a function $f(x)$ is said to be analytic at a point x_0 if it has an expansion in powers of $x - x_0$ in some neighborhood of x_0. Hence we have just shown that *a function analytic at a point x_0 is also analytic at every point of some neighborhood of x_0.*

   6. Prove the following generalization of the theorem on substitution of one power series into another: Suppose the functions $y = f(x)$ and $z = g(x)$ have power series expansions

$$y = f(x) = \sum_{n=0}^{\infty} a_n x^n, \quad z = g(x) = \sum_{n=0}^{\infty} b_n x^n, \tag{25}$$

both valid in the interval $(-R, R)$, while the function of two variables $\varphi(y, z)$ has a power series expansion

$$\varphi(y, z) = \sum_{k,m=0}^{\infty} h_{km} y^k z^m \tag{26}$$

in two variables (see *Rams.*, Sec. 10), valid for $|y| < \varrho$, $|z| < \varrho$. Then, if $|a_0| < \varrho$, $|b_0| < \varrho$, the composite function $\varphi(f(x), g(x))$ is defined in a neighborhood of $x = 0$, in which it has the expansion in powers of x obtained by replacing y and z in (26) by the series (25), multiplying the results of raising these series to the appropriate powers, and finally combining coefficients of like powers of x.

14. Division of Power Series

The theorem on substitution of one power series into another can be used to solve the problem of *division of power series*. Let

$$\sum_{n=0}^{\infty} a_n x^n = a_0 + a_1 x + a_2 x^2 + \cdots + a_n x^n + \cdots \quad (1)$$

be a power series with *nonzero* constant term a_0, and suppose we write (1) in the form

$$a_0 \left(1 + \frac{a_1}{a_0} x + \frac{a_2}{a_0} x^2 + \cdots + \frac{a_n}{a_0} x^n + \cdots \right) = a_0 (1 + y),$$

where

$$y = \frac{a_1}{a_0} x + \frac{a_2}{a_0} x^2 + \cdots + \frac{a_n}{a_0} x^n + \cdots .$$

Then

$$\frac{1}{a_0 + a_1 x + \cdots + a_n x^n + \cdots} = \frac{1}{a_0} \frac{1}{1 + y}$$

$$= \frac{1}{a_0} (1 - y + y^2 - \cdots + (-1)^m y^m + \cdots),$$

where the last series plays the role of the series (2) in the theorem on p. 97, with $\varrho = 1$. It follows from the theorem that

$$\frac{1}{a_0 + a_1 x + \cdots + a_n x^n + \cdots} = c_0 + c_1 x + \cdots + c_n x^n + \cdots,$$

at least for sufficiently small $|x|$, e.g., for x satisfying the inequality

$$\left| \frac{a_1}{a_0} \right| |x| + \left| \frac{a_2}{a_0} \right| |x|^2 + \cdots + \left| \frac{a_n}{a_0} \right| |x|^n + \cdots < 1.$$

Now let

$$\sum_{n=0}^{\infty} b_n x^n = b_0 + b_1 x + b_2 x^2 + \cdots + b_n x^n + \cdots$$

be another power series, with a nonzero radius of convergence. Then the ratio

$$\frac{b_0 + b_1 x + \cdots + b_n x^n + \cdots}{a_0 + a_1 x + \cdots + a_n x^n + \cdots} \tag{2}$$

can be replaced by the product

$$(b_0 + b_1 x + \cdots + b_n x^n + \cdots)(c_0 + c_1 x + \cdots + c_n x^n + \cdots)$$

for sufficiently small $|x|$, and hence can itself be represented as some power series

$$\sum_{n=0}^{\infty} d_n x^n = d_0 + d_1 x + d_2 x^2 + \cdots + d_n x^n + \cdots. \tag{3}$$

The coefficients of this series are most easily determined by the method of undetermined coefficients, starting from the identity

$$(a_0 + a_1 x + \cdots + a_n x^n + \cdots)(d_0 + d_1 x + \cdots + d_n x^n + \cdots)$$
$$= b_0 + b_1 x + \cdots + b_n x^n + \cdots,$$

where the coefficients $a_0, a_1, \ldots$ and $b_0, b_1, \ldots$ are regarded as known. Multiplying out the series on the left and then equating coefficients of like powers of x in both sides of the resulting identity, we get

$$a_0 d_0 = b_0, \ a_0 d_1 + a_1 d_0 = b_1, \ a_0 d_2 + a_1 d_1 + a_2 d_0 = b_2, \ldots,$$

$$a_0 d_n + a_1 d_{n-1} + \cdots + a_{n-1} d_1 + a_n d_0 = b_n, \ldots \tag{4}$$

Since the coefficient a_0 is nonzero, by assumption, the first of these equations gives

$$d_0 = \frac{b_0}{a_0};$$

then the second gives

$$d_1 = \frac{b_1 - a_1 d_0}{a_0} = \frac{a_0 b_1 - a_1 b_0}{a_0^2},$$

and so on. In general, once the coefficients $d_0, d_1, \ldots, d_{n-1}$ have been found, we can determine d_n from the $(n + 1)$st of the equations (4), in which d_n is the only remaining unknown. Thus (4) uniquely determines all the coefficients of (3) *recursively.*

Example 1. Find the first few terms of the ratio

$$\frac{x}{\ln \dfrac{1}{1 - x}} = \frac{x}{x + \dfrac{x^2}{2} + \dfrac{x^3}{3} + \dfrac{x^4}{4} + \cdots}$$

$$= \frac{1}{1 + \dfrac{x}{2} + \dfrac{x^2}{3} + \dfrac{x^3}{4} + \cdots}.$$

Solution. In this case, (4) takes the form

$$d_0 = 1, \qquad d_1 + \frac{1}{2} d_0 = 0, \qquad d_2 + \frac{1}{2} d_1 + \frac{1}{3} d_0 = 0,$$

$$d_3 + \frac{1}{2} d_2 + \frac{1}{3} d_1 + \frac{1}{4} d_0 = 0, \ldots,$$

and hence

$$d_0 = 1, \qquad d_1 = -\frac{1}{2}, \qquad d_2 = -\frac{1}{12}, \qquad d_3 = -\frac{1}{24}, \ldots$$

It follows that

$$\frac{x}{\ln \dfrac{1}{1 - x}} = 1 - \frac{1}{2} x - \frac{1}{12} x^2 - \frac{1}{24} x^3 - \cdots.$$

Example 2. Expand $\tan x$ in a neighborhood of zero, regarding $\tan x$ as the ratio of the functions $\sin x$ and $\cos x$ with known power series expansions.

Solution. The existence of such an expansion follows from the above theory. Moreover, the expansion contains only odd powers of x, since $\tan x$ is an odd function. Suppose we write the coefficient of x^{2n-1} in the required expansion in the form

$$\frac{T_n}{(2n-1)!}.$$

Then

$$\tan x = \frac{\sin x}{\cos x} = \sum_{n=1}^{\infty} \frac{T_n}{(2n-1)!} x^{2n-1},$$

and hence

$$\sum_{n=1}^{\infty} T_n \frac{x^{2n-1}}{(2n-1)!} \sum_{n=0}^{\infty} (-1)^n \frac{x^{2n}}{(2n)!} = \sum_{n=1}^{\infty} (-1)^{n-1} \frac{x^{2n-1}}{(2n-1)!},$$

$$(5)$$

after replacing $\sin x$ and $\cos x$ by their known power series expansions (*Ruds.*, p. 114). Clearly, $T_1 = 1$. To determine the remaining numbers T_n, we equate the coefficients of x^{2n-1} in both sides of (5). This gives

$$\frac{T_n}{(2n-1)!} - \frac{T_{n-1}}{(2n-3)!}\frac{1}{2!} + \frac{T_{n-2}}{(2n-5)!}\frac{1}{4!} - \cdots$$

$$= (-1)^{n-1}\frac{1}{(2n-1)!}$$

($n = 2, 3, \ldots$), or after multiplying by $(2n-1)!$,

$$T_n - C_2^{2n-1} T_{n-1} + C_4^{2n-1} T_{n-2} - \cdots = (-1)^{n-1},$$

where, as always, C_m^n denotes the binomial coefficient

$$\frac{n!}{m!\,(n-m)!}.$$

Since the numbers C_k^{2n-1} are all integers, so are all the numbers T_n. The values of the first few numbers T_n are easily found to be

$$T_1 = 1, \quad T_2 = 2, \quad T_3 = 16, \quad T_4 = 272,$$

$$T_5 = 7936, \ldots,$$

so that

$$\tan x = x + \frac{1}{3} x^3 + \frac{2}{15} x^5 + \frac{17}{315} x^7 + \frac{62}{2835} x^9 + \cdots. \tag{6}$$

Another way of calculating the coefficients in the series (6) will be given in Example 4, where we will also find the interval of convergence of (6).

Example 3. According to the above theory, the function

$$\frac{x}{e^x - 1} = \frac{x}{x + \dfrac{x^2}{2!} + \cdots + \dfrac{x^n}{n!} + \cdots}$$

$$= \frac{1}{1 + \dfrac{x}{2!} + \cdots + \dfrac{x^{n-1}}{n!} + \cdots}$$

has a power series expansion of the form

$$\frac{x}{e^x - 1} = 1 + \sum_{n=1}^{\infty} \frac{\beta_n}{n!} x^n, \tag{7}$$

at least for sufficiently small $|x|$, where we write the coefficients in the form $\beta_n/n!$ for convenience in subsequent calculations. As we will see below, this expansion has important applications. Starting from the formula

$$\left(1 + \frac{x}{2!} + \frac{x^2}{3!} + \cdots + \frac{x^{n-1}}{n!} + \cdots \right) \times$$

$$\times \left(1 + \frac{\beta_1}{1!} + \frac{\beta_2}{2!} + \cdots + \frac{\beta_n}{n!} x^n + \cdots \right) = 1,$$

we equate the coefficient of x^n to zero. This gives

$$\frac{1}{n!1!}\beta_n + \frac{1}{(n-1)!2!}\beta_{n-1} + \cdots$$

$$+ \frac{1}{(n-k+1)!k!}\beta_{n-k+1} + \cdots + \frac{1}{1!n!}\beta_1 + \frac{1}{(n+1)!} = 0$$

$$(8)$$

$(n = 1, 2, \ldots)$, or

$$C_1^{n+1}\beta_n + C_2^{n+1}\beta_{n-1} + \cdots + C_k^{n+1}\beta_{n+1-k} + \cdots$$

$$+ C_n^{n+1}\beta_1 + 1 = 0,$$

after multiplying by $(n + 1)!$. Exploiting the resemblance with the binomial theorem, we can write the last equation symbolically as

$$(\beta + 1)^{n+1} - \beta^{n+1} = 0 \quad (n = 1, 2, \ldots),$$

provided that after raising $\beta + 1$ to the power $n + 1$ by the usual rule and cancelling the leading term, we then replace the power β^k by the coefficient β_k. Thus we arrive at the following system of equations for determining the numbers β_k:

$$2\beta_1 + 1 = 0, \quad 3\beta_2 + 3\beta_1 + 1 = 0, \quad 4\beta_3 + 6\beta_2 + 4\beta_1 + 1 = 0,$$

$$5\beta_4 + 10\beta_3 + 10\beta_2 + 5\beta_1 + 1 = 0, \ldots$$

Solving these equations successively, we get

$$\beta_1 = -\frac{1}{2}, \qquad \beta_2 = \frac{1}{6}, \qquad \beta_3 = 0, \qquad \beta_4 = -\frac{1}{30},$$

$$\beta_5 = 0, \qquad \beta_6 = \frac{1}{42}, \qquad \beta_7 = 0, \qquad \beta_8 = -\frac{1}{30},$$

$$\beta_9 = 0, \qquad \beta_{10} = \frac{5}{66}, \qquad \beta_{11} = 0, \qquad \beta_{12} = -\frac{691}{2730},$$

$$\beta_{13} = 0, \qquad \beta_{14} = \frac{7}{6}, \qquad \beta_{15} = 0, \qquad \beta_{16} = -\frac{3617}{510},$$

$$\beta_{17} = 0, \quad \beta_{18} = \frac{43{,}867}{798}, \quad \beta_{19} = 0, \quad \beta_{20} = -\frac{174{,}611}{330}, \ldots$$

Since the numbers β_k satisfy linear equations with integral coefficients, they are all *rational*. It is easy to see that all the numbers β_k with odd indices vanish (except the first). In fact, transposing the term $-x/2$ to the left-hand side of (7), we get the even function

$$\frac{x}{e^x - 1} + \frac{x}{2} = \frac{x}{2} \frac{e^x + 1}{e^x - 1} = \frac{x}{2} \frac{e^{x/2} + e^{-x/2}}{e^{x/2} - e^{-x/2}} = \frac{x}{2} \coth \frac{x}{2}. \quad (9)$$

But the power series expansion

$$1 + \sum_{n=2}^{\infty} \frac{\beta_n}{n!} x^n \quad (10)$$

of the function (9) cannot contain odd powers of x.

We now write the numbers β_k with even indices in the form

$$\beta_{2n} = (-1)^{n-1} B_n, \quad (11)$$

where, as will be seen in a moment, the numbers B_n are all positive. Then we have

$$B_1 = \frac{1}{6}, \quad B_2 = \frac{1}{30}, \quad B_3 = \frac{1}{42}, \quad B_4 = \frac{1}{30},$$

$$B_5 = \frac{5}{66}, \quad B_6 = \frac{691}{2730}, \quad B_7 = \frac{7}{6}, \quad B_8 = \frac{3617}{510},$$

$$B_9 = \frac{43{,}867}{798}, \quad B_{10} = \frac{174{,}611}{330}, \ldots \quad (12)$$

The numbers B_n are called the *Bernoulli numbers* (after J. Bernoulli, who first introduced them), and play an important role in many problems of analysis. Using (11) and changing x to $2x$

in (9) and (10), we finally get

$$x \coth x = 1 + \frac{2^2 B_1}{2!} x^2 - \frac{2^4 B_2}{4!} x^4 + \cdots$$

$$+ (-1)^{n-1} \frac{2^{2n} B_n}{(2n)!} x^{2n} + \cdots$$

$$= 1 + \sum_{n=1}^{\infty} (-1)^{n-1} \frac{2^{2n} B_n}{(2n)!} x^{2n}. \tag{13}$$

It will be recalled from Problem 1, p. 96 that

$$\pi x \coth \pi x = 1 + 2 \sum_{n=1}^{\infty} (-1)^{n-1} s_{2n} x^{2n}, \tag{14}$$

where

$$s_{2n} = \sum_{m=1}^{\infty} \frac{1}{m^{2n}}. \tag{15}$$

Replacing x by πx in (13), we have

$$\pi x \coth \pi x = 1 + \sum_{n=1}^{\infty} (-1)^{n-1} \frac{(2\pi)^{2n} B_n}{(2n)!} x^{2n}. \tag{16}$$

Since the two expansions (14) and (16) must coincide, we have

$$B_n = \frac{2(2n)!}{(2\pi)^{2n}} s_{2n}. \tag{17}$$

In particular, (17) shows that the numbers B_n are all positive. Since $s_{2n} \to 1$ as $n \to \infty$, it follows from (17) that the Bernoulli numbers B_n approach infinity as $n \to \infty$, but nonmonotonically and very "capriciously," as is apparent from (12). According to (17), the sum s_{2n} is given by

$$s_{2n} = \sum_{m=1}^{\infty} \frac{1}{m^{2n}} = \frac{(2\pi)^{2n}}{2(2n)!} B_n. \tag{18}$$

In particular,

$$s_2 = \sum_{m=1}^{\infty} \frac{1}{m^2} = \frac{\pi^2}{6}, \qquad s_4 = \sum_{m=1}^{\infty} \frac{1}{m^4} = \frac{\pi^2}{90}.$$

Example 4. According to Example 3, p. 94, the function $\pi x \cot \pi x$ has the expansion

$$\pi x \cot \pi x = 1 - 2 \sum_{n=1}^{\infty} s_{2n} x^{2n}, \tag{19}$$

where s_{2n} again denotes the sum (15). Replacing πx by x and using (18), we find that

$$x \cot x = 1 - \sum_{n=1}^{\infty} \frac{2^{2n} B_n}{(2n)!} x^{2n}. \tag{20}$$

It will be recalled that the expansion (19) holds for $|x| < 1$, and hence the expansion (20) holds for $|x| < \pi$. But as $x \to \pm \pi$, the left-hand side of (20) approaches infinity. Therefore the series on the right cannot converge for $x = \pm \pi$, and hence certainly not for $x > \pi$. It follows that the radius of convergence of the series (20) is precisely π. Moreover, it is clear that the radius of convergence of the series (14) is also π, while that of the original series (7) is 2π.

Using the identity

$$\tan x = \cot x - 2 \cot 2x,$$

we see that (20) implies the following expansion for $\tan x$:

$$\tan x = \sum_{n=1}^{\infty} \frac{2^{2n} (2^{2n} - 1)}{(2n)!} B_n x^{2n-1}. \tag{21}$$

This series is identical with the expansion

$$\tan x = \sum_{n=1}^{\infty} \frac{T_n}{(2n - 1)!} x^{2n-1}$$

derived in Example 2, but the form (21) is preferable because of the familiarity of the Bernoulli numbers and the fact that they have been extensively tabulated. It is clear from the way it was derived that the series (21) has radius of convergence $\pi/2$.

PROBLEMS

1. Use the Cauchy–Hadamard theorem (*Ruds.*, p. 70) to prove that the radius of convergence R of the series (20) equals π.

Hint. Let ϱ_m be the mth root of the absolute value of the coefficient of x^m in (20). Then $\varrho_{2n-1} = 0$, while

$$\varrho_{2n} = \sqrt[2n]{\frac{2^{2n}B_n}{(2n)!}} = \sqrt[2n]{\frac{2^{2n}}{(2n)!}\frac{2\,(2n)!}{(2\pi)^{2n}}\,s_{2n}} = \frac{1}{\pi}\,\sqrt[2n]{2s_{2n}}\,,$$

so that

$$\varrho = \overline{\lim_{m \to \infty}}\,\varrho_m = \frac{1}{\pi}, \qquad R = \frac{1}{\varrho} = \pi.$$

2. Prove that

$$\ln\frac{\sin x}{x} = -\sum_{n=1}^{\infty}\frac{2^{2n}B_n}{(2n)!}\frac{x^{2n}}{2n} \qquad (|x| < \pi).$$

Hint. Note that

$$\frac{d}{dx}\ln\frac{\sin x}{x} = \frac{\cos x}{\sin x} - \frac{1}{x} = \frac{1}{x}(x\cot x - 1)$$

$$= -\sum_{n=1}^{\infty}\frac{2^{2n}B_n}{(2n)!}x^{2n-1}.$$

Now integrate term by term.

3. Prove that

$$\ln\cos x = -\sum_{n=1}^{\infty}\frac{2^{2n}(2^{2n}-1)\,B_n}{(2n)!}\frac{x^{2n}}{2n} \qquad \left(|x| < \frac{\pi}{2}\right).$$

Hint. Integrate (21) term by term.

4. Expand

$$\ln\frac{\tan x}{x}$$

in a neighborhood of the origin.

5. Sum the divergent series

$$\sum_{n=0}^{\infty} (-1)^n (n + 1)^k$$

by the Poisson–Abel method (*Rams.*, Sec. 16).

Hint. Consider the series

$$\sum_{n=0}^{\infty} (-1)^n e^{-(n+1)t}, \qquad (22)$$

where $t > 0$ and hence $0 < e^{-t} < 1$. Summing (22) as a geometric series, we get

$$\sum_{n=0}^{\infty} (-1)^n e^{-(n+1)t} = \frac{e^{-t}}{1 + e^{-t}} = \frac{1}{e^t + 1} = \frac{1}{e^t - 1} - \frac{2}{e^{2t} - 1}$$

$$= \frac{1}{t} \frac{t}{e^t - 1} - \frac{1}{t} \frac{2t}{e^{2t} - 1},$$

or

$$\sum_{n=0}^{\infty} (-1)^n e^{-(n+1)t} = -\sum_{n=1}^{\infty} \frac{(2^n - 1)\beta_n}{n!} t^{n-1}. \qquad (23)$$

Now differentiate both sides of (23) k times with respect to t (why is this justified?), obtaining

$$\sum_{n=0}^{\infty} (-1)^n (n + 1)^k e^{-(n+1)t}$$

$$= (-1)^{k+1} \sum_{n=k+1}^{\infty} \frac{(2^n - 1)\beta_n}{n!} (n - 1)(n - 2) \cdots (n - k) t^{n-k-1}.$$

Now let $t \to 0$ or equivalently $x \to 1$. Then the left-hand side approaches the generalized sum $A^{(k)}$ (in the Poisson–Abel sense), while the right-hand side reduces to the constant term

$$(-1)^{k+1} \frac{(2^{k+1} - 1)\beta_{k+1}}{k + 1}.$$

Therefore

$$A^{(2m)} = 0, \qquad A^{(2m-1)} = (-1)^{m-1} \frac{2^{2m} - 1}{2m} B_m \quad (m \geqslant 1).$$

Comment. The series (22) is also summable by the generalized Cesàro method of order k, to the same sum $A^{(k)}$ (see *Rams.*, Prob. 3, p. 119 and Theorem 3, p. 116).

15. Solution of Equations by Using Power Series

Given an equation of the form

$$F(x, y) = 0, \tag{1}$$

we now consider another version of the problem of determining the variable y as a function of x (cf. Sec. 9):

THEOREM. *Suppose the function $F(x, y)$ has a power series expansion in two variables $x - x_0$ and $y - y_0$ in a neighborhood of the point (x_0, y_0), where the constant term is zero and the coefficient of $y - y_0$ is nonzero, or equivalently, where*

$$F(x_0, y_0) = 0, \qquad F_y(x_0, y_0) \neq 0.$$

Then the function $y = y(x)$ defined by (1) in a neighborhood of x_0 also has a power series expansion in the variable $x - x_0$ near x_0.[4] In other words, if the function $F(x, y)$ figuring in the left-hand side of (1) is analytic at the point (x_0, y_0), then the function $y = y(x)$ defined by (1) is analytic at the point x_0.[5]

Proof. Without loss of generality, we can set $x_0 = y_0 = 0$, since this is equivalent to choosing the differences $x - x_0$ and $y - y_0$ as new variables, while retaining the old notation. Then, transposing the term containing the first power of y to the left-hand side and dividing by its coefficient, we can write the given

---

4. The existence of $y = y(x)$ in a neighborhood of x_0 follows from the implicit function theorem (p. 79).

5. Recall the remark on p. 50 and the comment to Problem 11, p. 55. Thus here we are talking not only about the existence and computation of the implicitly defined function $y = y(x)$, as in Sec. 9, but also about its *analytic representation.*

equation (1) in the form

$$y = c_{10}x + c_{20}x^2 + c_{11}xy + c_{02}y^2 + c_{30}x^3 + c_{21}x^2y$$
$$+ c_{12}xy^2 + c_{03}y^3 + \cdots. \tag{2}$$

We now look for a power series expansion of y in the form

$$y = a_1x + a_2x^2 + a_3x^3 + \cdots + a_nx^n + \cdots. \tag{3}$$

First we note that if such an expansion exists in a neighborhood of zero, then its coefficients are uniquely determined by equation (2). In fact, *assuming that* (3) *exists* and replacing y by (3) in (2), we get

$$a_1x + a_2x^2 + a_3x^3 + \cdots$$
$$= c_{10}x + c_{20}x^2 + c_{11}x(a_1x + a_2x^2 + a_3x^3 + \cdots)$$
$$+ c_{02}(a_1x + a_2x^2 + \cdots)^2 + c_{30}x^3 + c_{21}x^2(a_1x + \cdots)$$
$$+ c_{12}x(a_1x + \cdots)^2 + c_{03}(a_1x + \cdots)^3 + \cdots. \tag{2'}$$

It follows from the theorem on substitution of one power series into another (p. 97) that we can carry out all the operations on the right for sufficiently small $|x|$, afterwards combining coefficients of like powers of x. Equating coefficients of like powers of x. Equating coefficients of like powers of x in the resulting equation (cf. Theorem 3, p. 44), we arrive at the (infinite!) system of equations

$$a_1 = c_{10}, \qquad a_2 = c_{20} + c_{11}a_1 + c_{02}a_1^2,$$

$$\tag{4}$$

$$a_3 = c_{11}a_2 + 2c_{02}a_1a_2 + c_{30} + c_{21}a_1 + c_{12}a_1^2 + c_{03}a_1^3, \ldots$$

in the unknown coefficients $a_1, a_2, \ldots, a_n, \ldots$ Since all the terms containing y in the right-hand side of (2) are of degree two or higher (i.e., contain either y multiplied by some power of x or a higher power of y itself), the coefficient of a_n in the nth equation of (3) can be expressed in terms of the coefficients $a_1, a_2, \ldots, a_{n-1}$ with lower indices (and known coefficients c_{ik}).

This guarantees the possibility of determining the coefficients *recursively*:

$$a_1 = c_{10}, \qquad a_2 = c_{20} + c_{11}c_{10} + c_{02}c_{10}^2,$$

$$a_3 = (c_{11} + 2c_{02}c_{10})(c_{20} + c_{11}c_{10} + c_{02}c_{10}^2) + c_{30}$$

$$+ c_{21}c_{10} + c_{12}c_{10}^2 + c_{03}c_{10}^3, \ldots \tag{5}$$

We now make an observation which will be needed later. Since there are no operations other than addition and multiplication involved in carrying out the operations on the coefficients a_i and c_{ik} in the right-hand side of (2′), the expressions on the right in (4) must be polynomials in these quantities, in fact polynomials with positive integral coefficients. But then the expressions on the right in (5) must also be polynomials in the c_{ik} with positive integral coefficients.

Next we form the series (3) with the coefficients a_i given by (5). This series clearly satisfies (2′) *formally*. If we could prove the convergence of this series for sufficiently small $|x|$, then the proof that the function $y = y(x)$ represented by the series satisfies (2) would be complete, since then the formulas (5) satisfied by the coefficients of the series would be entirely equivalent to (2′). Thus *the whole problem now reduces to proving that the series (3) with coefficients a_i given by (5) actually converges in some neighborhood of zero.*

To this end, we now consider, besides (2), an analogous series

$$y = \gamma_{10}x + \gamma_{20}x^2 + \gamma_{11}xy + \gamma_{02}y^2 + \gamma_{30}x^3 + \gamma_{21}x^2y$$

$$+ \gamma_{12}xy^2 + \gamma_{03}y^3 + \cdots, \tag{2″}$$

where the coefficients γ_{ik} are all *positive* and moreover satisfy the inequalities

$$|c_{ik}| \leqslant \gamma_{ik}. \tag{6}$$

For the series (2″) we construct the (for the time being, formal) series

$$y = \alpha_1 x + \alpha_2 x^2 + \alpha_3 x^3 + \cdots + \alpha_n x^n + \cdots, \tag{3′}$$

analogous to (3), where the coefficients α_i are given by the formulas

$$\alpha_1 = \gamma_{10}, \qquad \alpha_2 = \gamma_{20} + \gamma_{11}\gamma_{10} + \gamma_{02}\gamma_{10}^2,$$

$$\alpha_3 = (\gamma_{11} + 2\gamma_{02}\gamma_{10})(\gamma_{20} + \gamma_{11}\gamma_{10} + \gamma_{02}\gamma_{10}^2) + \gamma_{30}$$

$$+ \gamma_{21}\gamma_{10} + \gamma_{12}\gamma_{10}^2 + \gamma_{03}\gamma_{10}^3, \cdots \tag{5'}$$

which are the exact analogues of (5). The very structure of these formulas, as already noted, guarantees that the numbers α_i are all *positive*. Moreover, comparing (5) and (5'), we find that

$$|a_n| \leqslant \alpha_n \quad (n = 1, 2, \ldots), \tag{7}$$

after using (6). If we manage to choose positive coefficients γ_{ik} not only satisfying (6) but also such that the corresponding series (3') has a nonzero radius of convergence, then, because of (7), the same would be true of the series (3), and our theorem would be proved. Thus we now concern ourselves with the choice of the numbers γ_{ik}.

Let r and ϱ be positive numbers such that the double series

$$|c_{10}|\, r + |c_{20}|\, r^2 + |c_{11}|\, r\varrho + |c_{02}|\, \varrho^2 + \cdots$$

converges.[6] Then $|c_{ik}|\, r^i \varrho^k \to 0$ as $i, k \to \infty$, and in particular,

$$|c_{ik}|\, r^i \varrho^k \leqslant M$$

for some positive constant M, i.e.,

$$|c_{ik}| \leqslant \frac{M}{r^i \varrho^k}.$$

If we set

$$\gamma_{ik} = \frac{M}{r^i \varrho^k},$$

---

6. The existence of such numbers follows from the analyticity of $F(x, y)$ (cf. *Rams.*, Lemma, p. 54).

the series (3') becomes

$$y = \frac{M}{r} x + \frac{M}{r^2} x^2 + \frac{M}{r\varrho} xy + \frac{M}{\varrho^2} y^2 + \cdots$$

$$= \frac{M}{\left(1 - \dfrac{x}{r}\right)\left(1 - \dfrac{x}{\varrho}\right)} - M - \frac{M}{\varrho} y,$$

so that finally

$$y^2 - \frac{\varrho^2}{\varrho + M} y + \frac{M\varrho^2}{\varrho + M} \frac{x}{r - x} = 0. \tag{8}$$

Here we can actually find the function $y = y(x)$ satisfying (8). In fact, solving the quadratic equation (8), we get

$$y = \frac{\varrho^2}{2(\varrho + M)} \left[1 - \sqrt{1 - \frac{4M(\varrho + M)}{2} \frac{x}{r - x}} \right], \tag{9}$$

where we choose the minus sign in order to obtain the branch for which $y = 0$ when $x = 0$. Setting

$$r_1 = r \left(\frac{\varrho}{\varrho + 2M} \right)^2, \tag{10}$$

we can write (9) in the form

$$y = \frac{\varrho^2}{2(\varrho + M)} \left[1 - \left(1 - \frac{x}{r_1} \right)^{1/2} \left(1 - \frac{x}{r} \right)^{-1/2} \right],$$

from which it is already clear by using the binomial series (*Ruds.*, Sec. 18) that y can be expanded in powers of x provided that $|x| < r_1 < r$. Since this expansion must be identical with (3'), this proves the convergence of (3') and hence that of (3), at least for $|x| < r_1$, thereby completing the proof of the theorem. ∎

Remark 1. Note that the theorem merely guarantees the possibility of expanding y in powers of x (or, more generally, in powers of $x - x_0$) in a neighborhood of the point zero (or the point x_0). The determination of the exact interval of convergence of the expansion requires a separate investigation.

Remark 2. The remarkable method of proof just given is due to Cauchy. The essence of the method consists in replacing the given power series (in one or more variables) by more suitable "majorant" series, all of whose coefficients are positive and exceed the absolute values of the corresponding coefficients of the given series. This technique, called *the method of majorant series,* is often used in the theory of differential equations.

As a special case of the above theory, we now consider the problem of *inverting a power series.* Suppose the function $y = y(x)$ has an expansion in powers of $x - x_0$ in a neighborhood of the point x_0. Letting y_0 denote the constant term (equal to the value of y for $x = x_0$), we can write this expansion in the form

$$y - y_0 = a_1 (x - x_0) + a_2 (x - x_0)^2 + \cdots$$
$$+ a_n (x - x_0)^n + \cdots. \tag{11}$$

If $a_1 \neq 0$, then (11) determines x as a function of y which has an expansion in powers of $y - y_0$ in a neighborhood of the point y_0. Thus, if $y = y(x)$ is an analytic function of x at the point x_0, then the inverse function $x = x(y)$ is analytic at the corresponding point y_0. All this follows from the theorem just proved (with the roles of x and y suitably interchanged). Letting $x_0 = y_0 = 0$ for simplicity, we write the relation between x and y in the form

$$x = by + c_2 x^2 + c_3 x^3 + c_4 x^4 + \cdots,$$

guided by the example of (2). Then the coefficients of the required expansion

$$x = b_1 y + b_2 y^2 + b_3 y^3 + \cdots \tag{12}$$

are determined recursively by the equations

$$b_1 = b, \qquad b_2 = c_2 b_1^2, \qquad b_3 = 2c_2 b_1 b_2 + c_3 b_1^3,$$

$$b_4 = c_2 (2b_1 b_3 + b_2^2) + 3c_3 b_1^2 b_2 + c_4 b_1^4,$$

$$b_5 = 2c_2 (b_1 b_4 + b_2 b_3) + 3c_3 (b_1^2 b_3 + b_1 b_2^2)$$

$$+ 4c_4 b_1^3 b_2 + c_5 b_1^5, \dots \tag{13}$$

(give the details).

Example. Find the expansion of $x = \arc\sin y$ by inverting the power series

$$y = \sin x = x - \frac{1}{6} x^3 + \frac{1}{120} x^5 - \cdots. \tag{14}$$

Solution. We write

$$x = \arc\sin y = y + b_3 y^3 + b_5 y^5 + \cdots,$$

using the fact that $x = \arc\sin y$, like $y = \sin x$ itself, is odd and hence has no even powers of y in its expansion. In this case, the equations (13) give

$$b_1 = 1, \qquad b_3 = \frac{1}{6} b_1^3 = \frac{1}{6},$$

$$b_5 = \frac{1}{2} b_1^2 b_3 - \frac{1}{120} b_1^5 = \frac{3}{40}, \dots$$

so that

$$x = \arc\sin y = y + \frac{1}{6} y^3 + \frac{3}{40} y^3 + \cdots, \tag{15}$$

in keeping with formula (6), p. 60.

The region in which the inverse function is guaranteed to exist and have the expansion (12) can be deduced from our previous considerations, but this often leads to a greatly reduced region of applicability. In the present case, for example, writing (14) in the form

$$x = y + \frac{x^3}{6} - \frac{x^5}{120} + \cdots,$$

like (2), and confining ourselves to values of x and y satisfying the inequalities $|x| < \pi/2, |y| < 1$, i.e., choosing $\varrho = \pi/2, r = 1$, we get $M = 1$ (why?), and hence, by formula (10),

$$r_1 = \left(\frac{\dfrac{\pi}{2}}{\dfrac{\pi}{2} + 2} \right)^2 < 0.2,$$

whereas the true region of applicability of the formula (15) is the whole interval $[-1, 1]$ (cf. Problem 1, p. 60)!

Remark. It is interesting to note what happens in the problem of inverting a power series, when the condition $a_1 \neq 0$ is waived. Thus suppose $a_1 = 0$, $a_2 \neq 0$, say $a_2 > 0$. Then we have

$$y = a_2 x^2 + a_3 x^3 + a_4 x^4 + \cdots$$

near $x = 0$ (setting $x_0 = y_0 = 0$ for simplicity), so that $y > 0$. Letting $\sqrt{y}$ denote the positive square root, we have

$$\sqrt{y} = \sqrt{a_2 x^2 + a_3 x^3 + a_4 x^4 + \cdots}$$

$$= \pm x \sqrt{a_2} \sqrt{1 + \frac{a_3}{a_2} x + \frac{a_4}{a_2} x^2 + \cdots},$$

where the sign of $\pm$ is chosen to coincide with that of x. According to the theorem on p. 120, the last factor on the right has an expansion in x near $x = 0$, with constant term 1. Thus, finally, transposing $\pm$ to the left-hand side, we have

$$\pm \sqrt{y} = a_1' x + a_2' x^2 + \cdots,$$

where $a_1' = \sqrt{a_2} > 0$. By the argument on p. 125, with y replaced by $\pm\sqrt{y}$, we get two different expansions for x near $x = 0$, depending on the choice of sign:

$$x_1 = b_1 y^{1/2} + b_2 y + b_3 y^{3/2} + b_4 y^2 + \cdots > 0,$$

$$x_2 = -b_1 y^{1/2} + b_2 y - b_3 y^{3/2} + b_4 y^2 - \cdots < 0$$

$(b_1 = 1/\sqrt{a_2} > 0)$. It should be noted that the inverse function now has two branches, where each branch involves fractional rather than integral powers of x.

PROBLEM

Find the expansion

$$\ln (1 + y) = y - \frac{1}{2} y^2 + \frac{1}{3} y^3 - \frac{1}{4} y^4 + \frac{1}{5} y^5 - \cdots$$

by inverting the series

$$y = e^x - 1 = x + \frac{x^2}{2!} + \frac{x^3}{3!} + \cdots.$$

Hint. In this case,

$$b_1 = 1, \qquad b_2 = -\frac{1}{2} b_1^2 = -\frac{1}{2},$$

$$b_3 = -b_1 b_2 - \frac{1}{6} b_1^3 = \frac{1}{3},$$

$$b_4 = -\frac{1}{2} (2b_1 b_3 + b_2^2) - \frac{1}{2} b_1^2 b_2 - \frac{1}{24} b_1^4 = -\frac{1}{4},$$

$$b_5 = -(b_1 b_4 + b_2 b_3) - \frac{1}{2} (b_1^2 b_3 + b_1 b_2^2) - \frac{1}{6} b_1^3 b_2$$

$$- \frac{1}{120} b_1^5 = \frac{1}{5}, \ldots$$

16. Lagrange's Series

We now apply the theorem on p. 120 to the equation of the special form

$$y = a + x\varphi (y), \tag{1}$$

where the function $\varphi(y)$ is assumed to be analytic at the point $y = a$. Then, as we know, (1) determines y as a function of x, which is analytic at the point $x = 0$ and equal to a for $x = 0$. Moreover, suppose $u = f(y)$ is a function of y which is analytic at the point $y = a$. Then, replacing y by the indicated function of x makes u into a function of x, which is also analytic at $x = 0$. We now pose the problem of expanding u in powers of x, more exactly of finding convenient expressions for the coefficients of such an expansion.

First we note that if a is also variable, then (1) determines y as a function of *two* variables x and a, analytic at the point $(0, a)$.[7] Then the function u will also be a function of the same two variables x and a. Differentiating (1) with respect to x and a, we get

$$[1 - x\varphi'(y)] \frac{\partial y}{\partial x} = \varphi(y), \qquad [1 - x\varphi'(y)] \frac{\partial y}{\partial a} = 1.$$

It follows at once that

$$\frac{\partial y}{\partial x} = \varphi(y) \frac{\partial y}{\partial a}, \tag{2}$$

and, more generally, if $u = f(y)$, that

$$\frac{\partial u}{\partial x} = \varphi(y) \frac{\partial u}{\partial a}. \tag{2'}$$

On the other hand, given any function $F(y)$ which is differentiable with respect to y, we have

$$\frac{\partial}{\partial x} \left[F(y) \frac{\partial u}{\partial a} \right] = \frac{\partial}{\partial a} \left[F(y) \frac{\partial u}{\partial x} \right], \tag{3}$$

as can be verified by direct differentiation and consultation of (2) and (2').

---

7. In making this assertion, we assume that the theorem on p. 120 can be extended to the case where the original equation involves *three* variables, thereby implicitly defining one of these variables as a function of the other two.

Using the above results, we now prove the important formula

$$\frac{\partial^n u}{\partial x^n} = \frac{\partial^{n-1}}{\partial a^{n-1}} \left[\varphi^n(y) \frac{\partial u}{\partial a} \right], \tag{4}$$

where $\varphi^n(y)$ denotes the nth power of $\varphi(y)$. For $n = 1$, (3) reduces to (2'). Assuming that (4) holds for some value $n \geqslant 1$, we now prove its validity for $n + 1$. To this end, we differentiate (4) with respect to x, obtaining[8]

$$\frac{\partial^{n+1} u}{\partial x^{n+1}} = \frac{\partial^{n-1}}{\partial a^{n-1}} \frac{\partial}{\partial x} \left[\varphi^n(y) \frac{\partial u}{\partial a} \right]. \tag{5}$$

But by (2') and (3), we have

$$\frac{\partial}{\partial x} \left[\varphi^n(y) \frac{\partial u}{\partial a} \right] = \frac{\partial}{\partial a} \left[\varphi^n(y) \frac{\partial u}{\partial x} \right] = \frac{\partial}{\partial a} \left[\varphi^{n+1}(y) \frac{\partial u}{\partial a} \right]. \tag{6}$$

Substituting (6) into (5), we get

$$\frac{\partial^{n+1} u}{\partial x^{n+1}} = \frac{\partial^n}{\partial a^n} \left[\varphi^{n+1}(y) \frac{\partial u}{\partial a} \right],$$

thereby proving (4) by induction.

We can now return to the problem of expanding the function u in powers of x. For constant a, this expansion must be a Taylor series of the form

$$u = u_0 + x \left(\frac{\partial u}{\partial x} \right)_0 + \frac{x^2}{2!} \left(\frac{\partial^2 u}{\partial x^2} \right)_0 + \cdots + \frac{x^n}{n!} \left(\frac{\partial^n u}{\partial x^n} \right)_0 + \cdots \tag{7}$$

(cf. Theorem 9, p. 49), where the subscript zero shows that the function and its derivatives are evaluated at $x = 0$. But then y takes the value a, so that $u_0 = f(a)$, and hence

$$\left(\frac{\partial^n u}{\partial x^n} \right)_0 = \frac{d^{n-1}}{da^{n-1}} [\varphi^n(a) f'(a)], \tag{8}$$

---

8. To justify interchanging the order of differentiation, see e.g., R. A. Silverman, *op. cit.*, Theorem 12.1, p. 708.

by formula (4). Substituting (8) into (7), we arrive at the expansion

$$f(y) = f(a) + x\varphi(a)f'(a) + \frac{x^2}{2!}\frac{d}{da}[\varphi^2(a)f'(a)] + \cdots$$

$$+ \frac{x^n}{n!}\frac{d^{n-1}}{da^{n-1}}[\varphi^n(a)f'(a)] + \cdots, \tag{9}$$

known as *Lagrange's series*. This expansion is remarkable in that its coefficients are represented as explicit functions of a. If $f(y) \equiv y$, then, in particular, (9) reduces to

$$y = a + x\varphi(a) + \frac{x^2}{2!}\frac{d}{da}[\varphi^2(a)] + \cdots$$

$$+ \frac{x^n}{n!}\frac{d^{n-1}}{da^{n-1}}[\varphi^n(a)] + \cdots. \tag{9'}$$

Remark. There is a close connection between the problem considered here and the problem of inverting a power series. Suppose we write equation (1) in the form

$$x = \frac{y-a}{\varphi(y)} = b_0 + b_1(y-a) + b_2(y-a)^2 + \cdots,$$

assuming that $\varphi(a) \neq 0$. Then Lagrange's problem is equivalent to inverting this expansion in powers of $y - a$. Conversely, in the problem of inverting the power series

$$y = a_1 x + a_2 x^2 + a_3 x^3 + \cdots \quad (a_1 \neq 0), \tag{10}$$

suppose we write (10) in the form

$$y = x(a_1 + a_2 x + a_3 x^2 + \cdots),$$

denoting the sum of the series in parentheses by $\psi(x)$. Then we get an equation

$$x = y\frac{1}{\psi(x)}$$

of the type (1), where $a = 0$, $\varphi(x) = 1/\psi(x)$ and the roles of x and y are interchanged. This allows us to use formula (9′) to give the following general expression for the result of inverting (10):

$$x = y\,\frac{1}{\psi(0)} + \frac{y^2}{2!}\left[\frac{d}{dx}\,\frac{1}{\psi^2(x)}\right]_{x=0} + \cdots$$

$$+ \frac{y^n}{n!}\left[\frac{d^{n-1}}{dx^{n-1}}\,\frac{1}{\psi^n(x)}\right]_{x=0} + \cdots. \tag{11}$$

Example 1. Consider the equation

$$y = x\,(a + x)\quad (a \neq 0), \tag{12}$$

or

$$x = y\,\frac{1}{a + x}.$$

Since

$$\frac{d^{n-1}}{dx^{n-1}}\,\frac{1}{(a + x)^n} = \frac{(-1)^{n-1}\,n\,(n + 1)\cdots(2n - 2)}{(a + x)^{2n-1}},$$

it follows from (11) that

$$x = \frac{y}{a} - \frac{y^2}{a^3} + \cdots + (-1)^{n-1}\,\frac{(2n - 2)!}{(n - 1)!n!}\,\frac{y^n}{a^{2n-1}} + \cdots.$$

The same expansion can be obtained by solving the quadratic equation (12) for x, and choosing the branch which vanishes when $y = 0$ (give the details).

Example 2. Consider the equation

$$y = a + \frac{x}{y}, \tag{13}$$

which is of the type (1) with $(y) = 1/y$. Suppose

$$f(y) = \frac{1}{y^k}.$$

Then (9) gives

$$\frac{1}{y_k} = \frac{1}{a_k} - x \frac{k}{a^{k+2}} + \frac{x^2}{2!} \frac{k(k+3)}{a^{k+4}} - \frac{x^3}{3!} \frac{k(k+4)(k+5)}{a^{k+6}}$$

$$+ \frac{x^4}{4!} \frac{k(k+5)(k+6)(k+7)}{a^{k+8}} - \cdots. \tag{14}$$

Since (13) is equivalent to the quadratic equation

$$y^2 - ay - x = 0,$$

we have

$$y = \frac{a}{2} + \sqrt{\frac{a^2}{4} + x},$$

choosing the plus sign to make $y = a$ for $x = 0$. Thus (14) implies the expansion

$$\left(\frac{2}{1 + \sqrt{1+x}}\right)^k = 1 - k\frac{x}{4} + \frac{k(k+3)}{2!}\left(\frac{x}{4}\right)^2$$

$$- \frac{k(k+4)(k+5)}{3!}\left(\frac{x}{4}\right)^3 + \cdots,$$

if we set $a = 2$ and multiply by 2^k.

PROBLEMS

1. In theoretical astronomy an important role is played by the equation
$$E = M + \varepsilon \sin E,$$

called *Kepler's equation*, where E is the "eccentric anomaly" of a given planet, M its "mean anomaly" and ε the eccentricity of the planet's orbit. Use Lagrange's series to expand E in powers of ε, with coefficients depending on M.

Ans. $E = M + \varepsilon \sin M + \frac{\varepsilon^2}{2!} \frac{d}{dM} \sin^2 M + \cdots$

$$+ \frac{\varepsilon^n}{n!} \frac{d^{n-1}}{dM^{n-1}} \sin^n M + \cdots. \tag{15}$$

Comment. It can be shown that the radius of convergence of the series (15) is 0.6627...[9]

2. By applying Lagrange's series to the equation

$$y = x + \frac{\alpha}{2}(y^2 - 1),\tag{16}$$

prove that

$$\frac{2x - \alpha}{1 + \sqrt{1 - 2\alpha x + \alpha^2}} = 1 + \alpha\frac{1}{2}\frac{d(x^2 - 1)}{dx}$$

$$+ \alpha^2\frac{1}{2!\,2^2}\frac{d^2(x^2 - 1)^2}{dx^2} + \cdots$$

$$+ \alpha^n\frac{1}{n!\,2^n}\frac{d^n(x^2 - 1)^n}{dx^n} + \cdots.$$

$$(17)$$

Hint. Solving (16) and choosing the branch equal to x when $\alpha = 0$, we get

$$y = \frac{1 - \sqrt{1 - 2\alpha x + \alpha^2}}{\alpha} = \frac{2x - \alpha}{1 + \sqrt{1 - 2\alpha x + \alpha^2}}.$$

3. By differentiating (17), deduce the formula

$$P_n(x) = \frac{1}{2^n\,n!}\frac{d^n(x^2 - 1)^n}{dx^n}\quad (n = 0, 1, 2, \ldots)$$

for the Legendre polynomials (Example 6, p. 104).

Hint. The legitimacy of differentiating (17) term by term follows from the analyticity of y as a function of the two variables α and x.

---

9. See A. I. Markushevich, *Theory of Functions of a Complex Variable, Volume II*, Prentice-Hall, Inc., Englewood Cliffs, N.J. (1965), Example 3, p.99 and Problem 3.11, p.113.

CHAPTER 5

Enveloping and Asymptotic Series

17. Definitions and Examples

We have already encountered some important ways of defining the "generalized sum" of a divergent series (*Rams.*, Chap. 4), where, of course, the partial sums of the series are unsuitable for approximate evaluation of such a "sum". Now, however, we consider divergent series from an entirely different standpoint, i.e., it will be shown that under certain conditions and within certain limits, the partial sums can, in fact, serve as excellent approximations to some number "generated" by the series in an appropriate sense. To give the reader some preliminary idea of the practical importance of divergent series in approximate calculations, we need only mention that this method is used by astronomers to predict the position of heavenly bodies, with results of perfectly satisfactory accuracy.

The following two simple examples illustrate the ideas figuring in this chapter:

Example 1. Consider the logarithmic series

$$x - \frac{x^2}{2} + \frac{x^3}{3} - \cdots + (-1)^{n-1} \frac{x^n}{n} + \cdots, \tag{1}$$

which, as is well known (*Ruds.*, Example 5, p. 115) represents the function $\ln(1 + x)$ only in the interval $-1 < x \leqslant 1$. Outside this interval (e.g., for $x > 1$), the series diverges and hence is meaningless. However, the function $\ln(1 + x)$ continues to be related to the partial sums of the series (1) even outside this

135

interval, since

$$\ln(1 + x) = x - \frac{x^2}{2} + \frac{x^3}{3} - \cdots + (-1)^{n-1}\frac{x^n}{n} + r_n(x)$$

by Taylor's series, where the remainder $r_n(x)$ can be written in Lagrange's form (say) as

$$r_n(x) = (-1)^n \frac{1}{n+1} \frac{x^{n+1}}{(1+\theta x)^{n+1}}$$

$$= (-1)^n \theta_1 \frac{x^{n+1}}{n+1} \quad (0 < \theta, \theta_1 < 1). \tag{2}$$

It follows that *the remainder has the same sign as the first omitted term and a smaller absolute value than this term,* just as in the case of a convergent series of the Leibniz type (*Ruds.*, p. 74)! Thus there is a convenient estimate of the error (and even the sign of the error) made in replacing the value of $\ln(1 + x)$ for $x > 1$ by the first n terms of the divergent series (1). *But this is enough to enable us to use the first n terms to calculate* $\ln(1 + x)$ *approximately*!

Of course, for $0 < x \leqslant 1$, the error approaches zero as $n \to \infty$, and moreover, given any n, we have

$$\frac{r_n(x)}{x^n} \to 0$$

as $x \to 0$, a fact summarized by writing

$$r_n(x) = o(x^n).$$

Hence the larger n, the higher the order of smallness of the error (relative to x). Given any fixed $x > 1$, the first omitted term, which serves as an estimate of the error, approaches infinity as $n \to \infty$. Therefore it cannot be said that the error can be made arbitrarily small for a given value of x. On the other hand, as the estimate

$$|r_n(x)| < \frac{x^{n+1}}{n+1}$$

itself shows, the error can in fact be made *arbitrarily small for x sufficiently near 1!* Moreover, if x is fixed and near 1, the terms of the series first decrease in absolute value, even for $x > 1$, in fact as long as

$$\frac{\dfrac{x^{n+1}}{n+1}}{\dfrac{x^n}{n}} = \frac{n}{n+1} \, x < 1$$

or

$$n < \frac{1}{x-1},$$

but then begin to increase in absolute value. Hence, to get the "best approximation" to the number $\ln(1+x)$ for a given x, it is best to truncate the series (1) at the term with index

$$n = \left[\frac{1}{x-1}\right]$$

(cf. footnote 2, p. 5).

In the preceding example, the series (1) does in fact converge for $-1 < x \leqslant 1$. The next example is more instructive, in that it involves an everywhere divergent series.

Example 2. Consider the *convergent* series

$$F(x) = \sum_{k=1}^{\infty} \frac{c_k}{x+k} \quad (0 < c < 1, \ x > 0). \tag{3}$$

For $k < x$ we have

$$\frac{1}{x+k} = \frac{1}{x} - \frac{k}{x^2} + \frac{k^2}{x^3} - \frac{k^3}{x^4} + \cdots. \tag{4}$$

However, the series diverges if $x \geqslant k$. Nevertheless, we can substitute (4) *formally* into the series (3) defining the function $F(x)$, and then combine like terms, thereby obtaining the series

$$\frac{A_1}{x} + \frac{A_2}{x^2} + \cdots + \frac{A_n}{x^n} + \cdots, \tag{5}$$

where

$$A_n = (-1)^{n-1} \sum_{k=1}^{\infty} k^{n-1} c_k.$$

It is easy to see that the series defining the coefficients A_n all converge. But the series (5) clearly diverges, since

$$|A_n| \geq n^{n-1} c^n, \qquad \left|\frac{A_n}{x^n}\right| \geq \frac{n^{n-1} c^n}{x^n},$$

and the last expression approaches infinity as $n \to \infty$.

The sum of the first n terms of the divergent series (5) equals

$$S_n(x) = \sum_{v=1}^{n} \frac{A_v}{x^v} = \sum_{k=1}^{\infty} c_k \sum_{v=1}^{n} \frac{(-1)^{v-1} k^{v-1}}{x^v}$$

$$= \sum_{k=1}^{\infty} \left[1 + (-1)^{n+1} \frac{k^n}{x^n} \right] \frac{c_k}{x + k},$$

and hence the remainder equals

$$r_n(x) = F(x) - S_n(x) = (-1)^n \sum_{k=1}^{\infty} \frac{k^n c^k}{(x + k) x^n}.$$

Thus, in this case,

$$r_n(x) = \frac{\theta(-1)^n}{x^{n+1}} \sum_{k=1}^{\infty} k^n c^k = \theta \frac{A_{n+1}}{x^{n+1}} \qquad (0 < \theta < 1),$$

as in (2), so that we again encounter the characteristic feature of a series of the Leibniz type, although the series (5) actually diverges. Of course, setting $F(x)$ approximately equal to the partial sum $S_n(x)$ of this divergent series does not give arbitrary accuracy for fixed x, but on the other hand *we can actually obtain any desired accuracy for sufficiently large x.* In the present case, as in the preceding example, it is still true that the number of terms retained should be increased (in the interest of improving accuracy) only as long as the terms decrease in absolute value, i.e., as long as $|A_{n+1}/A_n| < x$.

It is obvious that for fixed x, the remainder term $r_n(x)$ approaches zero as $x \to \infty$. Moreover,

$$x^n r_n(x) = \frac{\theta A_{n+1}}{x} \to 0$$

as $x \to \infty$, or, more concisely,

$$r_n(x) = o\left(\frac{1}{x^n}\right).$$

Thus the order of smallness of $r_n(x)$ is higher than n (relative to $1/x$). The more terms of the divergent series (5) that we retain, in order to get an approximate representation of the function $F(x)$, the higher the order of smallness of this approximation as $x \to \infty$.

Guided by these examples, we now turn to some key definitions:

DEFINITION 1. *Suppose the partial sums of a numerical series*

$$\sum_{n=0}^{\infty} a_n = a_0 + a_1 + a_2 + \cdots + a_n + \cdots \tag{6}$$

are alternately smaller and larger than some number A, i.e., suppose the "remainders" r_n defined by the formula

$$A = a_0 + a_1 + \cdots + a_n + r_n \tag{7}$$

are of alternating sign (as n increases). Then the series (16) is said to **envelop** *the number A (or to be an* **enveloping series** *for A).*

The simple formula

$$r_n = a_{n+1} + r_{n+1}$$

shows that Definition 1 is equivalent to

DEFINITION 1′. *The series (6) is said to* **envelop** *the number A if*

a) *The series is alternating;*

b) *The remainder term in formula (7) is less than a_{n+1} in absolute value and has the same sign as a_n.*[1]

---

1. We continue to use the same definition in the case where these conditions hold only for sufficiently large n (say for $n \geqslant n_0 > 1$).

Thus the series (1) envelops $\ln(1 + x)$ for any $x > 0$, while the series (5) envelops the function $F(x)$ for any $x > 0$.

Remark 1. The property of an enveloping series stated in Definition 1′ often makes it a valuable tool for the approximate calculation of the number A, but obviously not every series enveloping a number A is suitable for this purpose.

Remark 2. Definitions 1 and 1′ have obvious analogues (give them) for the case where we have a functional series

$$\sum_{n=0}^{\infty} a_n(x) = a_0(x) + a_1(x) + \cdots + a_n(x) + \cdots \qquad (6')$$

instead of the numerical series (6), and some function $A(x)$ instead of the number A. Here it is assumed that the function $A(x)$ and the functions $a_n(x)$ are all defined in one and the same domain X.

We now give a new definition pertaining to the special case where the domain of definition X of the terms of the series (6′) has a finite or infinite limit point ω. This time the remainder $r_n(x)$ is defined by the formula

$$A(x) = a_0(x) + a_1(x) + \cdots + a_n(x) + r_n(x), \qquad (7')$$

analogous to (7).

DEFINITION 2. *The series* (6′) *is said to be an* **asymptotic expansion** *of the function $A(x)$* **near the point $x = \omega$** *if* [2]

$$\lim_{x \to \omega} \frac{r_n(x)}{a_n(x)} = 0 \qquad (8)$$

for every fixed n, a fact indicated by writing

$$A(x) \sim a_0(x) + a_1(x) + \cdots + a_n(x) + \cdots.$$

Since

$$r_n(x) = a_{n+1}(x) + r_{n+1}(x)$$

---

2. Here, of course, it is assumed that $a_n(x)$ does not vanish, at least for x close enough to ω.

and

$$\frac{r_n(x)}{a_n(x)} = \frac{a_{n+1}(x)}{a_n(x)} \left[1 + \frac{r_{n+1}(x)}{a_{n+1}(x)} \right],$$

it follows from (8) that

$$\lim_{x \to \omega} \frac{a_{n+1}(x)}{a_n(x)} = 0. \tag{9}$$

Moreover, if the series (6') envelops the function $A(x)$ and if (9) holds, then (6') is also an asymptotic expansion of $A(x)$ near the point $x = \omega$. In fact, in this case,

$$|r_n(x)| \leqslant |a_{n+1}(x)|,$$

and hence

$$\left| \frac{r_n(x)}{a_n(x)} \right| \leqslant \left| \frac{a_{n+1}(x)}{a_n(x)} \right|,$$

so that (9) immediately implies (8).

Thus both series (1) and (5) are asymptotic expansions of the corresponding functions, the first near $x = 0$, the second near $x = \infty$.

In what follows, we will as a rule be concerned with asymptotic expansions of the form

$$A(x) \sim \sum_{n=0}^{\infty} \frac{a_n}{x^n} = a_0 + \frac{a_1}{x} + \frac{a_2}{x^2} + \cdots + \frac{a_n}{x^n} + \cdots \tag{10}$$

near $x = \infty$. The meaning of (10) is simply that, given any fixed n,

$$r_n = o\left(\frac{1}{x^n} \right),$$

or, more explicitly, that

$$\lim_{x \to \infty} \left[A(x) - a_0 - \frac{a_1}{x} - \frac{a_2}{x^2} - \cdots - \frac{a_n}{x^n} \right] x^n = 0. \tag{11}$$

Thus for "large" x we have the approximate formula

$$A(x) \approx a_0 + \frac{a_1}{x} + \frac{a_2}{x^2} + \cdots + \frac{a_n}{x^n},$$

where (11) characterizes the "quality" of this approximation. Writing (11) in the form

$$\lim_{x \to \infty} \left[A(x) - a_0 - \frac{a_1}{x} - \frac{a_2}{x^2} - \cdots - \frac{a_{n-1}}{x^{n-1}} \right] x^n = a_n, \quad (11')$$

we see that the asymptotic expansion (10) of the function $A(x)$ is *unique*, provided of course that $A(x)$ has such an expansion in the first place. In fact, all the coefficients a_n can be uniquely determined from (11') recursively.

Remark. For convenience we will sometimes write

$$B(x) \sim \varphi(x) + \psi(x) \sum_{n=0}^{\infty} \frac{a_n}{x^n}, \quad (12)$$

where $B(x)$, $\varphi(x)$ and $\psi(x)$ are functions defined on X, interpreting (12) to mean that

$$\frac{B(x) - \varphi(x)}{\psi(x)} \sim \sum_{n=0}^{\infty} \frac{a_n}{x^n}.$$

PROBLEMS

1. Show that a divergent series can envelop infinitely many numbers.

Hint. The series

$$1 - 2 + 2 - 2 + 2 - \cdots,$$

with partial sums $1, -1, 1, -1, \ldots$ envelops every number in the interval $(-1, 1)$.

2. Show that distinct functions can have one and the same asymptotic expansion.

Hint. If $A(x)$ has an asymptotic expansion near $x = \infty$, then every function of the form $A(x) + Ce^{-x}$ has the same asymptotic expansion, since $e^{-x}x^n \to 0$ as $x \to \infty$.

18. Basic Properties of Asymptotic Expansions

In this section, an asymptotic expansion will always mean an expansion of the form[3]

$$A(x) \sim \sum_{n=0}^{\infty} \frac{a_n}{x^n} = a_0 + \frac{a_1}{x} + \frac{a_2}{x^2} + \cdots + \frac{a_n}{x^n} + \cdots.$$

We will assume that all functions under consideration are defined in a domain X with limit point $+\infty$.

THEOREM 1. *If*

$$A(x) \sim \sum_{n=0}^{\infty} \frac{a_n}{x^n}, \qquad B(x) \sim \sum_{n=0}^{\infty} \frac{b_n}{x^n}, \qquad (1)$$

then

$$A(x) \pm B(x) \sim \sum_{n=0}^{\infty} \frac{a_n \pm b_n}{x^n}.$$

Proof. An immediate consequence of the meaning of the symbol $\sim$. ∎

THEOREM 2. *If $A(x)$ and $B(x)$ have asymptotic expansions* (1), *then the asymptotic expansion of the product $A(x)B(x)$ can be obtained by formal multiplication of the expansions* (1), *in accordance with Cauchy's rule.*[4]

Proof. It follows from (1) that

$$A(x) = a_0 + \frac{a_1}{x} + \frac{a_2}{x^2} + \cdots + \frac{a_n}{x^n} + o\left(\frac{1}{x^n}\right),$$

$$B(x) = b_0 + \frac{b_1}{x} + \frac{b_2}{x^2} + \cdots + \frac{b_n}{x^n} + o\left(\frac{1}{x^n}\right)$$

---

3. The theory of asymptotic expansions was developed by H. Poincaré, who made important applications of such expansions to celestial mechanics and the theory of differential equations.

4. See *Rams.*, p. 16.

for arbitrary n. Therefore

$$A(x)\,B(x) = c_0 + \frac{c_1}{x} + \frac{c_2}{x^2} + \cdots + \frac{c_n}{x^n} + o\left(\frac{1}{x^n}\right), \qquad (2)$$

where

$$c_m = \sum_{i=0}^{m} a_i b_{m-i}.$$

But (2) is equivalent to

$$A(x)\,B(x) \sim \sum_{n=0}^{\infty} \frac{c_n}{x^n}. \qquad \blacksquare$$

Remark. If $B(x) = A(x)$, we get the asymptotic expansion of the square $[A(x)]^2$. The asymptotic expansion of the mth power $[A(x)]^m$ ($m = 3, 4, \ldots$) can be found in the same way.

THEOREM 3. *Let $F(y)$ be any function analytic at the point $y = 0$, i.e., with a power series expansion*

$$F(y) = \sum_{m=0}^{\infty} \beta_m y^m = \beta_0 + \beta_1 y + \beta_2 y^2 + \cdots + \beta_m y^m + \cdots$$

in a neighborhood of zero, and let $A(x)$ be any function with an asymptotic expansion of the form

$$A(x) \sim \frac{a_1}{x} + \frac{a_2}{x^2} + \cdots + \frac{a_n}{x^n} + \cdots \qquad (3)$$

with no constant term, so that $A(x) \to 0$ as $x \to \infty$. Then the function

$$F\,(A(x)) = \sum_{m=0}^{\infty} \beta_m\,[A(x)]^m \qquad (4)$$

has an asymptotic expansion which can be obtained by replacing every power $[A(x)]^m$ by its asymptotic expansion and afterwards combining coefficients of like powers of x.[5]

---

5. Note the similarity with the theorem on substitution of one power series into another (p. 97).

Proof. Clearly $F(A(x))$ is defined, at least for sufficiently large x. The function $F(y)$ clearly has a continuous (and hence bounded) derivative in a neighborhood of the point $y = 0$, and hence

$$|F(\bar{y}) - F(y)| \leqslant C |\bar{y} - y| \quad (C = \text{const})$$

for any two points y and $\bar{y}$ of the neighborhood. Let $A_n(x)$ denote the sum of the first n terms of the series (3):

$$A_n(x) = \frac{a_1}{x} + \frac{a_2}{x^2} + \cdots + \frac{a_n}{x^n}.$$

Both $A(x)$ and $A_n(x)$ lie in the indicated neighborhood for sufficiently large x and fixed n. Therefore

$$|x^n [F(A(x)) - F(A_n(x))| \leqslant Cx^n |A(x) - A_n(x)|$$

$$= Cx^n |r_n(x)| \to 0$$

as $x \to \infty$, and moreover

$$F(A(x)) = F(A_n(x)) + o\left(\frac{1}{x^n}\right). \tag{5}$$

On the other hand, by the theorem on p. 97,

$$F(A_n(x)) = \beta_0 + \sum_{m=1}^{\infty} \beta_m (A_n(x))^m = \beta_0 + \frac{\beta_1 a_1}{x} + \frac{\beta_1 a_2 + \beta_2 a_1^2}{x^2}$$

$$+ \frac{\beta_1 a_3 + 2\beta_2 a_1 a_2 + \beta_3 a_1^3}{x^2} + \cdots$$

$$+ \frac{\beta_1 a_n + \cdots + \beta_n a_1^n}{x^n} + o\left(\frac{1}{x^n}\right).$$

But the same formula also holds for the function $F(A(x))$, because of (5), i.e.,

$$F(A(x)) \sim \beta_0 + \frac{\beta_1 a_1}{x} + \frac{\beta_1 a_2 + \beta_2 a_1^2}{x^2}$$

$$+ \frac{\beta_1 a_3 + 2\beta_2 a_1 a_2 + \beta_3 a_1^3}{x^3} + \cdots + \frac{\beta_1 a_n + \cdots + \beta_n a_1^n}{x^n},$$

as asserted. ∎

Next we consider integration of asymptotic series:

THEOREM 4. *Let $A(x)$ be continuous in the interval $X = [c, +\infty)$, where it has the asymptotic expansion*

$$A(x) \sim \frac{a_2}{x^2} + \frac{a_3}{x^3} + \cdots + \frac{a_n}{x^n} + \cdots, \tag{6}$$

beginning with the term containing $1/x^2$. Then $A(x)$ has a finite integral from any $x \geqslant c$ to $+\infty$,[6] and this integral (regarded as a function of x) has the asymptotic expansion

$$\int_x^\infty A(x)\, dx \sim \frac{a_2}{x} + \frac{a_3}{2}\frac{1}{x^2} + \cdots + \frac{a_n}{n-1}\frac{1}{x^{n-1}} + \cdots, \tag{7}$$

obtained by integrating (6) term by term.

Proof. Let

$$A(x) = \sum_{k=2}^n \frac{a_k}{x^k}, \qquad r_n(x) = A(x) - A_n(x);$$

Given any $\varepsilon > 0$ and any fixed n, we have

$$x^n\, |r_n(x)| < \varepsilon \tag{8}$$

for sufficiently large x. If $\xi > x$, then

$$\int_x^\xi A(x)\, dx = \int_x^\xi A_n(x)\, dx + \int_x^\xi r_n(x)\, dx$$

$$= \sum_{k=2}^\infty \frac{a_k}{k-1}\left(\frac{1}{x^{k-1}} - \frac{1}{k-1}\right) + \int_x^\xi r_n(x)\, dx.$$

---

6. It will be recalled that the integral of the function $f(x)$ from c to $+\infty$ means the limit

$$\int_c^{+\infty} f(x)\, dx = \lim_{C \to +\infty} \int_c^C f(x)\, dx$$

(*Ruds.*, p. 43).

As $\xi \to \infty$, we get

$$\int_{x_i^1}^{\infty} A(x)\, dx = \frac{a_2}{1}\frac{1}{x} + \frac{a_3}{2}\frac{1}{x^2} + \cdots + \frac{a_n}{n-1}\frac{1}{x^{n-1}} + R_{n-1}(x),$$
$$(9)$$

where

$$R_{n-1}(x) = \lim_{\xi \to \infty} \int_x^{\xi} r_n(x)\, dx = \int_x^{\infty} r_n(x)\, dx.$$

But

$$\left| \int_x^{\xi} r_n(x)\, dx \right| \leq \int_x^{\xi} |r_n(x)|\, dx < \varepsilon \int_x^{\xi} \frac{dx}{x^n}$$

$$= \frac{\varepsilon}{n-1}\left(\frac{1}{x^{n-1}} - \frac{1}{\xi^{n-1}} \right),$$

because of (8). Hence, taking the limit as $\xi \to \infty$, we have

$$|R_{n-1}(x)| \leq \frac{\varepsilon}{x^{n-1}}$$

(for sufficiently large x), so that

$$\lim_{x \to \infty} x^{n-1} R_{n-1}(x) = 0. \qquad (10)$$

Comparing (9) and (10), we get the asymptotic expansion (7). ∎

PROBLEMS

1. Prove that if

$$A(x) \sim \frac{a_1}{x} + \frac{a_2}{x^2} + \cdots + \frac{a_n}{x^n} + \cdots,$$

then

$$e^{A(x)} \sim 1 + \frac{a_1}{x} + \left[\frac{a_2}{1!} + \frac{a_1^2}{2!} \right]\frac{1}{x^2} + \left[\frac{a_3}{1!} + \frac{2a_1 a_2}{2!} + \frac{a_1^3}{3!} \right]\frac{1}{x^3} + \cdots$$

$$+ \left[\frac{a_n}{1!} + \cdots + \frac{a_1^n}{n!} \right]\frac{1}{x^n} + \cdots.$$

2. State and prove the analogue for asymptotic series of the treatment of division of power series, given on pp. 109–111.

3. Why does Theorem 4 fail if $A(x)$ contains a term of the form

$$\frac{a_1}{x} \quad (a_1 \neq 0)?$$

4. Prove that term-by-term differentiation of an asymptotic series is in general not permissible.

Hint. If $F(x) = e^{-x} \sin e^x$, then

$$\lim_{x \to \infty} F(x) \, x^n = 0$$

for arbitrary n, so that $F(x) \sim 0$, i.e., $F(x)$ has an asymptotic expansion identically equal to zero. On the other hand, the derivative $F'(x) = e^{-x} \sin e^x + \cos e^x$ has no asymptotic expansion, since even the limit

$$\lim_{x \to \infty} F'(x)$$

fails to exist.

19. The Euler–Maclaurin Formula

Next we prove a result known as the *Euler–Maclaurin formula*, which plays an important role in analysis and, in particular, is often used to find enveloping and asymptotic expansions. Our starting point is Taylor's formula[7]

$$\Delta f(x_0) = f(x_0 + h) - f(x_0)$$

$$= hf'(x_0) + \frac{h^2}{2!} f''(x_0) + \cdots + \frac{h^m}{m!} f^{(m)}(x_0) + \varrho,$$

---

7. It is tacitly assumed that all derivatives appearing here and elsewhere exist and are continuous.

where the remainder term

$$\varrho = \frac{1}{m!} \int_{x_0}^{x_0+h} f^{(m+1)}(t)(x_0 + h - t)^m \, dt$$

$$= \int_0^h f^{(m+1)}(x_0 + h - z) \frac{z^m}{m!} \, dz$$

in the form of a definite integral (*Def. Int.*, Sec. 11). Suppose we choose f to be the functions

$$\frac{1}{h} \int_{x_0}^x f(t) \, dt, \; f(x), \; hf'(x), \; h^2 f''(x), \ldots, h^{m-2} f^{(m-2)}(x)$$

in turn, at the same time replacing m by

$$m, m - 1, m - 2, m - 3, \ldots, 1.$$

This gives the following system of m equations:

$$\frac{1}{h} \int_{x_0}^{x_0+h} f(t) \, dt$$
$$= f(x_0) + \frac{h}{2!} f'(x_0) + \frac{h^2}{3!} f''(x_0) + \cdots + \frac{h^{m-1}}{m!} f^{(m-1)}(x_0) + \varrho_0 \quad \Big| \; 1$$

$$\Delta f(x_0)$$
$$= \qquad\quad hf'(x_0) + \frac{h^2}{2!} f''(x_0) + \cdots + \frac{h^{m-1}}{(m-1)!} f^{(m-1)}(x_0) + \varrho_1 \quad \Big| \; A_1$$

$$h \Delta f'(x_0)$$
$$= \qquad\qquad\qquad\quad h^2 f''(x_0) + \cdots + \frac{h^{m-1}}{(m-2)!} f^{(m-1)}(x_0) + \varrho_2 \quad \Big| \; A_2$$

$$\cdots\cdots\cdots\cdots\cdots\cdots\cdots\cdots\cdots\cdots\cdots\cdots\cdots\cdots\cdots \quad \Big| \; \cdots\cdots$$

$$h^{m-2} \Delta f^{(m-2)}(x_0)$$
$$= \qquad\qquad\qquad\qquad\qquad\qquad \frac{h^{m-1}}{1!} f^{(m-1)}(x_0) + \varrho_{m-1} \quad \Big| \; A_{m-1}$$

We now exclude all derivatives from the right-hand sides of the equations of this system, adding to the first equation (multiplied by 1) the second equation multiplied by A_1, the third equation multiplied by A_2, etc., and finally the last (mth) equation multiplied by A_{m-1}, where the numbers $A_1, A_2, \ldots, A_{m-1}$ are chosen to make

$$\frac{1}{2!} + A_1 = 0, \qquad \frac{1}{3!} + \frac{1}{2!} A_1 + A_2 = 0, \ldots,$$

$$\frac{1}{m!} + \frac{1}{(m-1)!} A_1 + \frac{1}{(m-2)!} A_2 + \cdots + A_{m-1} = 0. \quad (1)$$

As a result, we get

$$f(x_0) = \frac{1}{h} \int_{x_0}^{x_0+h} f(t)\, dt + A_1\, \Delta f(x_0) + A_2 h\, \Delta f'(x_0) + \cdots$$

$$+ A_{m-1} h^{m-2}\, \Delta f^{(m-2)}(x_0) + r, \qquad (2$$

where

$$r = -\varrho_0 - A_1 \varrho_1 - A_2 \varrho_2 - \cdots - A_{m-1} \varrho_{m-1}$$

$$= -\frac{1}{h} \int_0^h f^{(m)}(x_0 + h - z) \left[\frac{z^m}{m!} + A_1 \frac{h z^{m-1}}{(m-1)!} \right.$$

$$\left. + A_2 \frac{h^2 z^{m-2}}{(m-2)!} + \cdots + A_{m-1} h^{m-1} z \right] dz,$$

or, more briefly,

$$r = -\frac{1}{h} \int_0^h f^{(m)}(x_0 + h - z)\, \varphi_m(z)\, dz, \qquad (3)$$

where

$$\varphi_m(z) = \frac{z^m}{m!} + A_1 \frac{h z^{m-1}}{(m-1)!} + A_2 \frac{h^2 z^{m-2}}{(m-2)!} + \cdots$$

$$+ A_{m-1} h^{m-1} z. \qquad (4)$$

Clearly, the system (1) uniquely determines each of the coefficients $A_1, A_2, \ldots, A_{m-1}$ in succession, independently of the choice of the function f and the numbers x_0 and h. Moreover, these numbers are just the coefficients $\beta_k/k!$ of the expansion of the function

$$\frac{x}{e^x - 1}$$

in powers of x, given by equation (7), p. 113, as can be seen at once by comparing the last of the equations (1) with equation (8), p. 14, with n replaced by $m - 1$. Recalling pp. 114–115, we find that

$$A_1 = \frac{\beta_1}{1!} = -\frac{1}{2},$$

$$A_{2p-1} = \frac{\beta_{2p-1}}{(2p-1)!} = 0 \quad \text{if} \quad p > 1, \qquad (5)$$

$$A_{2p} = \frac{\beta_{2p}}{(2p)!} = (-1)^{p-1} \frac{B_p}{(2p)!},$$

where B_p is the *pth Bernoulli number*.

Now suppose the function $f(x)$ is defined in a finite interval $[a, b]$. Let

$$h = \frac{b - a}{n},$$

where n is a positive integer, and choose x_0 to be the numbers

$$a, a + h, a + 2h, \ldots, a + (n - 1) h = b - h$$

in turn. Writing an equation like (2), with remainder (3), for every subinterval

$$[a + (i - 1) h, a + ih] \quad (i = 1, 2, \ldots, n),$$

and adding all the resulting equations, we get the *Euler–Maclaurin formula*

$$\sum_{i=1}^{n} f(a + (i - 1)) h \equiv \sum_{a}^{b} f(x) = \frac{1}{h} \int_{a}^{b} f(x) \, dx$$

$$+ A_1 [f(b)] - f(a)] + A_2 h [f'(b) - f'(a)] + \cdots$$

$$+ A_{m-1} h^{m-2} [f^{(m-2)}(b) - f^{(m-2)}(a)] + R, \qquad (6)$$

with remainder

$$R = -\frac{1}{h} \sum_{i=1}^{n} \int_{0}^{h} f^{(m)} (a + ih - z) \, \varphi_m(z) \, dz$$

$$\equiv \frac{1}{h} \sum_{a}^{b} \int_{0}^{h} f^{(m)} (x + h - z) \, \varphi_m(z) \, dz, \qquad (7)$$

where $\varphi_m(z)$ is given by (4) and m is any positive integer beginning with 2.

Next we prove some properties of the function $\varphi_m(z)$, as a preliminary to investigating the remainder (7). First of all, we observe that differentiation of (4) gives

$$\varphi'_m(z) = \varphi_{m-1}(z) + A_{m-1} h^{m-1}. \qquad (8)$$

Moreover,

$$\varphi_m(0) = 0, \quad \varphi_m(h) = 0 \quad (m = 2, 3, \ldots), \qquad (9)$$

where the first formula is obvious from (4) and the second follows from the last equation of the system (1).

LEMMA. *The function of even order $\varphi_{2k}(z)$ cannot take any value more than twice as z varies in the interval $[0, h]$.*

Proof. Assume the contrary. Then the derivative of $\varphi_{2k}(z)$, equal to

$$\varphi'_{2k}(z) = \varphi_{2k-1}(z)$$

(since $A_{2k-1} = 0$), vanishes at both end points of the interval $[0, h]$, and hence, by a slight extension of Rolle's theorem,[8]

---

8. P. P. Korovkin, *Differentiation*, Theorem 2.2, p. 42.

vanishes no less than twice inside this interval. But then, by the same theorem, the derivative

$$\varphi'_{2k-1}(z) = \varphi_{2k-2}(z) + A_{2k-2}h^{2k-2}$$

vanishes no less than three times inside the interval $[0, h]$, i.e., the function $\varphi_{2k-2}(z)$ takes the same value $-A_{2k-2}h^{2k-2}$ no less than three times inside $[0, h]$. Thus, progressively lowering the order of the function $\varphi_{2k}(z)$ by 2, we arrive at the conclusion that the function

$$\varphi_2(z) = \frac{1}{2} z^2 - \frac{h}{2} z$$

takes some value no less than three times, which is impossible, since $\varphi_2(z)$ is a polynomial of degree 2. This contradiction proves the theorem. ∎

COROLLARY. *The function $\varphi_{2k}(z)$ does not change sign in the interval $(0, h)$.*

Proof. By (9), $\varphi_{2k}(z)$ vanishes at the end points of $(0, h)$. Hence, by the lemma, $\varphi_{2k}(z)$ cannot vanish again in $(0, h)$. ∎

Remark 1. The sign of the polynomial $\varphi_{2k}(z)$ is easily determined. In fact, for small z (and hence, by the corollary, everywhere between 0 and h), $\varphi_{2k}(z)$ has the sign of the lowest-order term $A_{2k-2}h^{2k-2}z^2$ (note again that $A_{2k-1} = 0$), i.e., the sign $(-1)^k$, since

$$A_{2k-2} = (-1)^{k-2} \frac{B_{k-1}}{(2k-2)!}.$$

Remark 2. Thus each of the consecutive functions of even order $\varphi_{2k}(z)$ and $\varphi_{2k+2}(z)$ has a definite sign in $(0, h)$, but their signs are opposite.

Turning now to the remainder term R, given by (7), suppose m is an even number $m = 2k$, and assume that the derivatives $f^{(2k)}(z)$ and $f^{(2k+2)}(z)$ both have the same sign in the interval $[a, b]$. Then, integrating the expression for R twice by parts and taking account of (8) and (9), we successively get

$$R = -\frac{1}{h} \sum_a^b \int_0^h \varphi_{2k}(z) f^{(2k)}(x + h - z)\, dz$$

$$= \frac{1}{h} \sum_a^b \int_0^h [A_{2k}h^{2k} - \varphi'_{2k+1}(z)] f^{(2k)}(x + h - z)\, dz$$

$$= \frac{1}{h} A_{2k}h^{2k} \sum_a^b [f^{(2k-1)}(x + h) - f^{(2k-1)}(x)]$$

$$- \frac{1}{h} \sum_a^b \int_0^h \varphi_{2k+1}(z) f^{(2k+1)}(x + h - z)\, dz$$

$$= A_{2k}h^{2k-1} [f^{(2k-1)}(b) - f^{(2k-1)}(a)]$$

$$- \frac{1}{h} \sum_a^b \int_0^h \varphi'_{2k+2}(z) f^{(2k+1)}(x + h - z)\, dz$$

$$= A_{2k}h^{2k-1} [f^{(2k-1)}(b) - f^{(2k-1)}(a)]$$

$$- \frac{1}{h} \sum_a^b \int_0^h \varphi_{2k+2}(z) f^{(2k+2)}(x + h - z)\, dz.$$

Since the two underlined expressions have opposite signs, because of Remark 2 and our assumption about $f^{(2k)}(z)$ and $^{(2k+2)}(z)$, *the first of these expressions must have the same sign as f*

$$A_{2k}h^{2k-1} [f^{(2k-1)}(b) - f^{(2k-1)}(a)], \tag{10}$$

and at the same time must be less than (10) *in absolute value.* Thus, finally,

$$R = R_{2k} = \theta A_{2k}h^{2k-1} [f^{(2k-1)}(b) - f^{(2k-1)}(a)]$$

$$= \theta(-1)^{k-1} \frac{B_k}{(2k)!} h^{2k-1} [f^{(2k-1)}(b) - f^{(2k-1)}(a)]$$

$$(0 < \theta < 1). \tag{11}$$

Assuming now that *all* the derivatives of even order $f^{(2k)}(z)$ have the same sign in the interval $[a, b]$, and writing an infinite series instead of the finite sum (6), we get the *Euler–Maclaurin series*

$$\sum_a^b f(x) = \frac{1}{h} \int_a^b f(x)\, dx - \frac{1}{2} [f(b) - f(a)] + \frac{B_1}{2!} h\, [f'(b) - f'(a)]$$

$$- \frac{B_2}{4!} h^3\, [f'''(b) - f'''(a)] + \cdots$$

$$+ (-1)^{k-2} \frac{B_{k-1}}{(2k-2)!} h^{2k-3}\, [f^{(2k-3)}(b) - f^{(2k-3)}(a)]$$

$$+ (-1)^{k-1} \frac{B_k}{(2k)!} h^{2k-1} [f^{(2k-1)}(b) - f^{(2k-1)}(a)] + \cdots, \tag{12}$$

after using the expressions (5) for the coefficients A_m. This series *diverges* in general, so that the equality sign in (12) is written only provisionally. Moreover, the series is alternating, at least starting from the third term, because of our assumption about the derivatives $f^{(2k)}(z)$. Recalling (11) and Definition 1′, p. 139, we see that the series *envelops* the sum

$$\sum_a^b f(x) \tag{13}$$

appearing on the left. If we interchange this sum and the integral

$$\frac{1}{h} \int_a^b f(x)\, dx, \tag{14}$$

at the same time changing the signs of all the other terms, we get a series *enveloping* the integral (14). The partial sums of these series sometimes allow us to make very accurate calculations of the sum (13) from a knowledge of the integral (14), or of the integral (14) from a knowledge of the sum (13). Here, of course, a basic role is played by the fact that we know how to estimate the remainder term in advance.

We now give two examples illustrating the above considerations:

Example 1. Calculate the following sum of nine hundred (!) terms:

$$\sum_{i=100}^{i=999} \frac{1}{i} \equiv \sum_{100}^{1000} \frac{1}{x}.$$

Solution. Here we have

$$f(x) = \frac{1}{x}, \qquad a = 100, \qquad b = 1000, \qquad h = 1.$$

Since

$$f'(x) = -\frac{1}{x^2}, \qquad f''(x) = \frac{2}{x^3}, \qquad f'''(x) = -\frac{6}{x^4},$$

$$f^{(iv)}(x) = \frac{24}{x^5}, \qquad f^{(v)}(x) = -\frac{120}{x^6},$$

and in general,

$$f^{(2k)}(x) = \frac{(2k)!}{x^{2k+1}},$$

all the derivatives of even order have the same (positive) sign. Using the Euler–Maclaurin formula (6) to carry out the expansion up to the term containing f''', so that the remainder contains $f^{(v)}$, we obtain

$$\sum_{100}^{1000} \frac{1}{x} = \int_{100}^{1000} \frac{dx}{x} + \frac{1}{2}\left(\frac{1}{100} - \frac{1}{1000}\right) + \frac{1}{12}\left(\frac{1}{100^2} - \frac{1}{1000^2}\right)$$

$$- \frac{6}{720}\left(\frac{1}{100^4} - \frac{1}{1000^4}\right) + \theta\,\frac{12}{3024}\left(\frac{1}{100^6} - \frac{1}{1000^6}\right),$$

where $0 < \theta < 1$. Carrying out the numerical calculations (give the details[9]), we find that

$$\sum_{100}^{1000} \frac{1}{x} = 2.30709334291072$$

to within an error of 10^{-14}!

Example 2. Use the Euler–Maclaurin formula to evaluate the integral

$$\int_0^1 \frac{dx}{1 + x} = \ln 2.$$

Solution. This time we have

$$f(x) = \frac{1}{1 + x}, \qquad a = 0, \qquad b = 1.$$

Let

$$h = \frac{1}{10} \quad (n = 10).$$

Since

$$f'(x) = -\frac{1}{(1 + x)^2}, \qquad f''(x) = \frac{2}{(1 + x)^3},$$

$$f'''(x) = -\frac{6}{(1 + x)^4},$$

$$f^{(iv)}(x) = \frac{24}{(1 + x)^5}, \qquad f^{(v)}(x) = -\frac{120}{(1 + x)^6},$$

and in general,

$$f^{(2k)}(x) = \frac{(2k)!}{(1 + x)^{2k+1}},$$

---

9. In particular, note that

$$\int_{100}^{1000} \frac{dx}{x} = \ln 10.$$

so that all the derivatives of even order have the same (positive) sign. Using the modified Euler–Maclaurin formula (with the sum and integral interchanged), again truncated at the term containing f''', we get

$$\int_0^1 \frac{dx}{1+x} = \frac{1}{10} + \frac{1}{11} + \frac{1}{12} + \frac{1}{13} + \frac{1}{14} + \frac{1}{15} + \frac{1}{16} + \frac{1}{17} + \frac{1}{18}$$

$$+ \frac{1}{19} - \frac{1}{20}\left(1 - \frac{1}{2}\right) - \frac{1}{1200}\left(1 - \frac{1}{4}\right)$$

$$+ \frac{6}{7,200,000}\left(1 - \frac{1}{16}\right) - \theta \frac{12}{3,024,000,000}\left(1 - \frac{1}{64}\right)$$

$$(0 < \theta < 1).$$

Carrying out the numerical calculations (give the details), we find that

$$\int_0^1 \frac{dx}{1+x} = \ln 2 = 0.69314718$$

to within an accuracy of 10^{-8}.

PROBLEMS

1. Use the Euler–Maclaurin formula to prove that

$$\sum_{i=0}^{\infty} \frac{1}{(a+i)^2} = \frac{1}{a} + \frac{1}{2}\frac{1}{a^2} + B_1 \frac{1}{a^3} - B_2 \frac{1}{a^5} + B_3 \frac{1}{a^7} - \cdots$$

$$+ (-1)^{k-2} B_{k-1} \frac{1}{a^{2k-1}}$$

$$+ \theta (-1)^{k-1} B_k \frac{1}{a^{2k+1}} \quad (0 \leqslant \theta \leqslant 1), \qquad (15)$$

where a is a positive integer.

Hint. It follows from (6) and (7) with

$$f(x) = \frac{1}{x^2}, \quad h = 1, \quad b = a + nh$$

that

$$\sum_{i=0}^{n-1} \frac{1}{(a+i)^2} = -\left[\frac{1}{a+n} - \frac{1}{a}\right] - \frac{1}{2}\left[\frac{1}{(a+n)^2} - \frac{1}{a^2}\right]$$

$$- B_1\left[\frac{1}{(a+n)^3} - \frac{1}{a^3}\right]$$

$$+ B_2\left[\frac{1}{(a+n)^5} - \frac{1}{a^5}\right] - \cdots$$

$$- (-1)^{k-2} B_{k-1}\left[\frac{1}{(a+n)^{2k-1}} - \frac{1}{a^{2k-1}}\right]$$

$$- \theta_n(-1)^{k-1} B_k\left[\frac{1}{(a+n)^{2k+1}} - \frac{1}{a^{2k+1}}\right]$$

$$(0 < \theta_n < 1).$$

Now take the limit as $n \to \infty$, with a and k held fixed, justifying the fact that θ_n then approaches some limit θ in the interval $[0, 1]$.

2. Prove that
$$\pi^2 = 9.86960440108935862$$

to within an error of 10^{-17}.

Hint. Choosing $a = 10$ and $k = 10$ in (15), we get

$$\pi^2 = 6\sum_{i=1}^{\infty} \frac{1}{i^2} = 6\sum_{i=1}^{9} \frac{1}{i^2} + \frac{6}{10} + \frac{3}{100} + \frac{1}{1000} - \frac{1}{5 \cdot 10^5}$$

$$+ \frac{1}{7 \cdot 10^7} - \frac{1}{5 \cdot 10^9} + \frac{5}{11 \cdot 10^{11}} - \frac{691}{455 \cdot 10^{13}}$$

$$+ \frac{7}{10^{15}} - \frac{3617}{85 \cdot 10^{17}} + \frac{43,867}{133 \cdot 10^{19}}$$

$$- \theta \frac{174,611}{55 \cdot 10^{21}} \quad (0 \leqslant \theta \leqslant 1)$$

(recall p. 116).

Comment. Thus we have evaluated the sum π^2 of a slowly converging series with great accuracy, by using the partial sum of a *divergent* series enveloping π^2. To achieve the same accuracy by using the original convergent series, we would need more than a billion terms!

20. Another Form of the Euler–Maclaurin Formula

Returning to formulas (6) and (7), p. 152, we now assume that the function $f(z)$ has derivatives of all orders in the infinite interval $[a, +\infty)$, which satisfy the following conditions:

1) The derivatives of even order $f^{(2k)}(z)$ all have the same sign in $[a, +\infty)$;
2) The derivatives of odd order $f^{(2k-1)}(z)$ all approach zero as $z \to +\infty$.

As before, let m be an even number $m = 2k$. Holding the numbers a and k fixed, while regarding $b = a + nh$ (and n itself) as variable, we then write the remainder R in the form

$$
\begin{aligned}
R = &-\frac{1}{h} \sum_{i=1}^{\infty} \int_0^h \varphi_{2k}(z)\, f^{(2k)}\,(a + ih - z)\, dz \\
&+ \frac{1}{h} \sum_{i=n+1}^{\infty} \int_0^h \varphi_{2k}(z)\, f^{(2k)}\,(a + ih - z)\, dz \\
\equiv &-\frac{1}{h} \sum_{a}^{\infty} \int_0^h \varphi_{2k}(z)\, f^{(2k)}\,(x + h - z)\, dz \\
&+ \frac{1}{h} \sum_{b}^{\infty} \int_0^h \varphi_{2k}(z)\, f^{(2k)}\,(x + h - z)\, dz.
\end{aligned}
$$

We now combine the first of these sums with all the terms of formula (6) containing a to form a single constant

$$
\begin{aligned}
C_k = &-A_1 f_1(a) - A_2 h f'(a) - \cdots - A_{2k-2} h^{2k-3} f^{(2k-3)}(a) \\
&- \frac{1}{h} \sum_{a}^{\infty} \int_0^h \varphi_{2k}(z) f^{(2k)}\,(x + h - z)\, dz
\end{aligned}
$$

(recall that $A_{2k-1} = 0$), which is explicitly independent of b. We can then write formula (6), p. 152 in the form

$$\sum_a^b f(x) = C_k + \frac{1}{h} \int_a^b f(x)\, dx + A_1 f(b) + A_2 h f'(b) + \cdots$$

$$+ A_{2k-2} h^{2k-3} f^{(2k-3)}(b) + R', \qquad (1)$$

where

$$R' = \frac{1}{h} \sum_{i=n+1}^{\infty} \int_0^h \varphi_{2k}(z) f^{(2k)}(a + ih - z)\, dz$$

$$= \frac{1}{h} \sum_{i=1}^{\infty} \int_0^h \varphi_{2k}(z) f^{(2k)}(b + ih - z)\, dz$$

$$\equiv \frac{1}{h} \sum_b^{\infty} \int_0^h \varphi_{2k}(z) f^{(2k)}(x + h - z)\, dz.$$

To justify the above transformation, we need only convince ourselves of the *convergence* of the infinite series involved. We begin with the series

$$\frac{1}{h} \sum_a^b \int_0^h \varphi_{2k}(z) f^{(2k)}(x + h - z)\, dz.$$

It follows from the considerations on p. 154 that

$$0 < \frac{\dfrac{1}{h} \sum_{i=1}^{n-1} \int_0^h \varphi_{2k}(z) f^{(2k)}(a + ih - z)\, dz}{A_{2k} h^{2k-1} [f^{(2k-1)}(a) - f^{(2k-1)}(a + nh)]} < 1.$$

Using the corollary on p. 153 and condition 1) above, we see that the terms in the numerator of this fraction *all have the same sign*, i.e., the sign of the denominator. Hence, taking the limit as $n \to \infty$ and recalling condition 2), we conclude that the series

$$\frac{1}{h} \sum_a^{\infty} \int_0^h \varphi_{2k}(z) f^{(2k)}(x + h - z)\, dz$$

$$\equiv \frac{1}{h} \sum_{i=1}^{\infty} \int_0^h \varphi_{2k}(z) f^{(2k)}(a + ih - z)\, dz$$

converges, and moreover that the sum of the series has the same sign as the expression

$$A_{2k}h^{2k-1}f^{(2k-1)}(a),$$

which it does not exceed in absolute value. Similarly, by the same argument with *a* changed to *b*, we conclude that the series

$$\frac{1}{h}\sum_{b}^{\infty}\int_{0}^{h}\varphi_{2k}(z)f^{(2k)}(x+h-z)\,dz$$

$$\equiv \frac{1}{h}\sum_{i=1}^{\infty}\int_{0}^{h}\varphi_{2k}(z)f^{(2k)}(b+ih-z)\,dz$$

also converges, and moreover that the sum of the series has the same sign as the epxression

$$A_{2k}h^{2k-1}f^{(2k-1)}(b),$$

which it does not exceed in absolute value. Thus, finally, we have not only verified the convergence of the infinite series in question, but we have incidentally proved that the remainder R' in formula (1) can be written as

$$R' = \theta A_{2k}h^{2k-1}f^{(2k-1)}(b) = \theta(-1)^{k-1}\frac{B_k}{(2k)!}h^{2k-1}f^{(2k-1)}(b)$$

$$(0 < \theta < 1). \tag{2}$$

By the very way it is formed, the possibility of the constant C_k in formula (1) depending on k is hardly precluded. However, curiously enough, C_k *is in fact independent of* k. To see this, we need only compare formulas (1) and (2) with the same formulas written for $k = 1$, i.e.,

$$\sum_{a}^{b}f(x) = C_1 + \frac{1}{h}\int_{a}^{b}f(x)\,dx + A_1f(b) + \bar{R}',$$

where
$$\bar{R}' = \bar{\theta} A_2 h f'(b) \quad (0 < \bar{\theta} < 1).$$

As a result, we get

$$C_1 + \bar{\theta} A_2 h f'(b) = C_k + A_2 h f'(b) + \cdots + A_{2k-2} h^{2k-3} f^{(2k-3)}(b)$$

$$+ \theta A_{2k} h^{2k-1} f^{(2k-1)}(b).$$

Taking the limit as $b \to +\infty$ and using condition 2), p. 160, we find that
$$C_k = C_1 = C.$$

The constant C, known as the *Euler–Maclaurin constant* for the function $f(x)$, depends not only on $f(x)$, but also on a and h.

Remark 1. Since $<$ must be replaced by $\leqslant$ in taking limits in an inequality, it seems that we should write $0 \leqslant \theta \leqslant 1$ in formula (2). However, the case $\theta = 0$ is immediately excluded, since the sum of an infinite series whose terms are all of the same sign cannot vanish. Moreover, if $\theta = 1$, then, increasing the index k by 1 in formula (1), we would have $R' = 0$, which is impossible, for the reason just given. Thus we actually have $0 < \theta < 1$, as written in (2).

Remark 2. Writing an infinite series instead of (1), we get the following version of the Euler–Maclaurin series

$$\sum_a^b f(x) = C + \frac{1}{h} \int_a^b f(x)\,dx - \frac{1}{2} f(b) + \frac{B_1}{2!} h f'(b)$$

$$- \frac{B_2}{4!} h^3 f'''(b) + \cdots$$

$$+ (-1)^{k-2} \frac{B_{k-1}}{(2k-2)!} h^{2k-3} f^{(2k-3)}(b)$$

$$+ (-1)^{k-1} \frac{B_k}{(2k)!} h^{2k-1} f^{(2k-1)}(b) + \cdots \qquad (3)$$

(here the equality sign is written only provisionally!). By condition 1), all the derivatives $f^{(2k-1)}(b)$ change in the same direction as b increases, and hence they all have the same sign, since they approach zero as $b \to +\infty$, by condition 2). Together with formula (2), this shows that the Euler–Maclaurin series *envelops* the sum

$$\sum_a^b f(x)$$

on the left (recall Definition 1′, p. 139).

Remark 3. To determine the constant C figuring in (3), choose some $b > a$ for which the sum and integral are easily calculated. Then we get the series

$$C = \sum_a^b f(x) - \frac{1}{h} \int_0^h f(x) \, dx + \frac{1}{2} f(b) - \frac{B_1}{2!} h f'(b)$$

$$+ \frac{B_2}{4!} h^3 f'''(b) - \cdots$$

enveloping C, from which an approximate value of C can be calculated in many cases.

Example. As an important application of the Euler–Maclaurin series, we now calculate the quantity

$$\ln(n!) = \ln n + \sum_{i=1}^{n-1} \ln i.$$

Choosing

$$a = 1, \qquad h = 1, \qquad b = n$$

(with $n - 1$ instead of n), let

$$f(x) = \ln x.$$

Then

$$f^{(m)}(x) = (-1)^{m-1} \frac{(m-1)!}{x^m},$$

so that conditions 1) and 2), p. 160 are satisfied. Using (3), we arrive at the asymptotic expansion

$$\ln(n!) = \ln n + \sum_{1}^{n} f(x) \sim C + \left(n + \frac{1}{2}\right) \ln n - n$$

$$+ \frac{B_1}{1 \cdot 2} \frac{1}{n} - \frac{B_2}{3 \cdot 4} \frac{1}{n^2} + \cdots$$

$$+ (-1)^{k-1} \frac{B_k}{(2k-1)\,2k} \frac{1}{n^{2k-1}} + \cdots, \tag{4}$$

called *Stirling's series*, where C includes an extra constant of integration equal to 1. The absolute value of the general term of the series (4) is

$$\frac{s_{2k}}{2\pi^2 n} \frac{(2k-2)!}{(2\pi n)^{2k-2}}, \tag{5}$$

where

$$s_{2k} = \sum_{m=1}^{\infty} \frac{1}{m^{2k}} = \frac{(2\pi)^{2k}}{2\,(2k)!} B_k,$$

as on p. 116. Hence (4) diverges, since (5) approaches infinity as $k \to \infty$.[10]

Terminating the series (4) at the term containing B_{k-1} and including a remainder term, we get

$$\ln(n!) = C + \left(n + \frac{1}{2}\right) \ln n - n + \frac{B_1}{1 \cdot 2} \frac{1}{n} - \frac{B_2}{3 \cdot 4} \frac{1}{n^3} + \cdots$$

$$+ (-1)^{k-2} \frac{B_{k-1}}{(2k-3)\,(2k-2)} \frac{1}{n^{2k-3}}$$

$$+ \theta(-1)^{k-1} \frac{B_k}{(2k-1)\,2k} \frac{1}{n^{2k-1}} \quad (0 < \theta < 1). \tag{6}$$

---

10. To show this, use formula (7) below and the fact that $s_{2k} \to 1$ as $k \to \infty$.

In the special case $k = 1$, this reduces to

$$\ln (n!) = C + \left(n + \frac{1}{2}\right) \ln n - n + \frac{\theta}{12n},$$

or

$$n! = e^C \sqrt{n} \left(\frac{n}{e}\right)^n e^{\theta/12n}, \qquad (7)$$

after taking exponentials. This is just *Stirling's formula*, proved by another method in *Ruds.*, Sec. 19, where it was shown that

$$e^C = \sqrt{2\pi}.$$

Thus the heretofore undetermined constant C is just $\frac{1}{2} \ln 2\pi$.

PROBLEMS

1. Prove that

$$n! \sim \sqrt{2\pi n} \left(\frac{n}{e}\right)^n \times$$

$$\times \left\{1 + \frac{1}{12n} + \frac{1}{288n^2} - \frac{139}{51,840n^3} - \frac{571}{2,488,320n^4} + \cdots\right\}$$

Hint. Use (4) and Theorem 3, p. 144.

2. Using (6), calculate $\ln (100!)$ to eight decimal places. *Ans.* 363.73937555...

Index